Ulrich Lüttge   Physiological Ecology of Tropical Plants

Springer

Berlin
Heidelberg
New York
Barcelona
Budapest
Hong Kong
London
Milan
Paris
Santa Clara
Singapore
Tokyo

Ulrich Lüttge

Physiological Ecology of Tropical Plants

With 235 Figures and 42 Tables

 Springer

Professor Dr. Ulrich Lüttge

Technische Hochschule Darmstadt
Institut für Botanik
Schnittspahnstraße 3–5
D-64287 Darmstadt
Germany

Cover photo showing the fringe of a rainforest on Sierrania Parú, Guayana Highlands, Venezuela; 1250 m a.s.l., 04°25'N, 65°32'W.

ISBN 3-540-61161-4 Springer-Verlag Berlin Heidelberg New York

Library of Congress Cataloging-in-Publication Data

Lüttge, Ulrich.
  Physiological ecology of tropical plants / Ulrich Lüttge.
    p. cm.
  Includes bibliographical reference and index.
  ISBN 3-540-61161-4
  1. Tropical plants – Ecophysiology. 2. Plant ecophysiology
– Tropics. I. Title.
QK936.L88 1997
581.7'0913 – dc21

Cover photograph: Prof. U. Lüttge, Darmstadt
Typesetting: Mitterweger Werksatz GmbH, D-68723 Plankstadt · Germany
SPIN 10530316   31/3137   5 4 3 2 1 0 - Printed on acid-free papier

Preface

Recently, in many countries, particularly in the industrialized parts of the world, there has been an upsurge in public opinion concerned about the alarming rate of destruction of tropical ecosystems by man, and particularly the continuing elimination of tropical rainforest. The response to the public concern has led to a bourgeoning of popular literature. More dispassionate scientific books are often devoted to special topics of tropical ecology, e.g. biotopes such as rainforests or savannas, ecologically defined groups of plants such as mangroves, epiphytes or succulent plants, and special taxa such as palms. The ecological approach of this scientific literature is predominantly floristic (and faunal) description and analysis of the diversity in associations, biotopes and ecosystems.

However, the development of modern experimental technology, which is increasingly well adapted to the use in field work in the tropics, is also allowing more and more detailed ecophysiological studies. Observations in the field lead to delineation of precise problems for studies in laboratories, growth chambers and phytotrons. The results of such work are built into hypotheses, whose ecological significance in turn is tested again in the field. This fruitful ecophysiological interplay between work in the field and in the laboratory leads to an increasing understanding of physiological, biochemical and molecular bases of ecological adaptations. Phenotypic physiological plasticity is important in mechanisms of ecological adaptations and may also be involved in mechanisms of generation and maintenance of floristic and faunal diversity in ecosystems.

I am therefore convinced that ecology must be studied at various levels of complexity and integration, namely the phytogeographical, ecosystem or biotope level, also called the synecological level, the whole organism level, also called the autecological level, and the cellular, organelle, membrane and molecular levels. These various levels may, in fact, not be so much distinguished by their respective degree of complexity. One may consider them as fractals. With each additional magnification, i.e. on each level, similar problems of complexity with integration of subsystems, feedback and non-linear behaviour will be encountered. Therefore, the distinction between levels does not appear to be basically conceptual and is more a matter of scaling. Studies with ecological relevance must not remain isolated within the individual levels. In addition to the interplay between work in the field and in the laboratory, there must be continuous feedback and feedforward within and between the levels of different scaling. Thus, although the aim is to progress towards levels at finer scales as far as possible, I think we must also refer to the levels at larger scales in order to put mechanisms of ecological adaptations into context.

Alexander von Humboldt was the first to recognize the relations between physiognomy of plants and the environment. In this vein, it seemed appropriate to begin the various chapters and sections of this book with descriptions of the physiognomy of biotopes and plants and to deduce the ecophysiological problems from them. This may also help to motivate readers, depending on their individual starting points, either to be carried on from the experienced environment to more abstract levels of understanding, or to consider the function of molecular, biochemical or physiological units in relation to the performance of plants in habitats.

It is the aim of this book to cover plants of all major tropical ecosystems. In the tropics we encounter biotopes with vast expanses. Thus, I felt that a brief record of current trends in large scale sensing and diagnosis would be useful (Chap. 2). The largest and most dominating tropical biomes are forests (Chap. 3) and savannas (Chap. 7). Ecophysiology of tropical plants is, in general, still a limited field, which is in its early stages of development, and in some areas knowledge is still poor. In limited areas, however, progress is already quite advanced. For example much work is available on epiphytes, which therefore have been given their own chapter (Chap. 4) although they are important parts of tropical forests. Mangroves are very specific tropical forests and also are treated in a separate chapter (Chap. 5). Salinas, inselbergs and páramos are very characteristic tropical environments and are covered each in a short chapter (Chap. 6, 8 and 9). Although plants of dry and arid habitats are discussed in various chapters, genuine deserts are excluded. First, the major deserts lie outside the tropics. Second, very much ecophysiological work has already been performed on desert plants, and this would go much beyond the scope of this book.

Physiognomy is depicted with photographs, most of which were taken during my own excursions in the tropics. Ecophysiological work on tropical plants was compiled from the literature whereever available and I also refer to studies of my own group. Some original publications are cited when necessary, but when possible and appropriate preference was given to quoting summarizing and reviewing works, because it is the aim of this book to give a general overview rather than specific interpretations of specialized research contributions.

Abundant illustrations with drawings and tables are used to elucidate ecophysiological relations. It was, moreover intended to write a simple and readily flowing text, which is easy to read. In fact, it was aimed to make this book useful for a wide audience interested in the tropical environment. Thus, while writing I found that, for some readers, basic knowledge behind ideas, results and hypotheses presented would need at least brief repetition and explanation. In order to avoid interruption of the text, this was separated into boxes, which the reader may consult or overlook, depending on the individual level of understanding.

I am indebted to former students and postdoctoral fellows in my laboratory and friends and colleagues who work in various tropical countries and made it possible for me to become acquainted with tropical environments and to perform ecophysiological field research in the tropics. Most of this occurred in the Caribbean, in Venezuela and Brazil. Unavoidably, this led to a noticeable South American bias of the book. This may be of some disadvantage in the selection of the concrete physiognomic descriptions, but in the more abstract ecophysiological context it may be less important.

There is much goodwill in industrialized countries for an understanding of the tropics, but much more knowledge is needed. Many efforts are being made, and there is much research in the developing tropical countries, but more encouragement is needed. I hope this book, may make some contribution, if humble, towards both. I have written this book in English, which is not my mother tongue, but the scientific lingua franca to address people in the different parts of the world. In striving to build bridges, Alexander von Humboldt is chosen as a mentor for this book, and citations from his lucid descriptions of his journey to South America are used here and there as a guide.

I owe particular thanks to Professor Dr. Dr. h.c. mult. Hubert Ziegler for his encouragement in publishing this text, and also to him and Professor Dr. Erwin Beck for reading some of the chapters. With great empathy, Professor Dr. Howard Griffiths went through the painstaking work of correcting my English, and with a great deal of sensitivity, being an expert in tropical plant ecophysiology himself, he made invaluable contributions much beyond establishing linguistic discipline. I thank the students who attended my courses and lectures on tropical plant ecophysiology and served as guinea pigs for the development of this text. My own research in the tropics and exchange with colleagues in tropical countries were supported by the Deutsche Forschungsgemeinschaft (DFG), the Alexander-von-Humboldt-Stiftung, the Deutscher Akademischer Austauschdienst (DAAD), the Volkswagenstiftung and the Körber European Science Award, which are most gratefully acknowledged. A great many thanks are due to Ms. Barbara Reinhards, who took care of an almost endless succession of different versions of the manuscript with never-exhausted patience, and Ms. Doris Schäfer and Ms. Rosel Heger who handled the line drawings and the halftone photographs, respectively.

Darmstadt, February 1997 ULRICH LÜTTGE

Contents

CHAPTER 1

Introduction

1.1
Historical Background of Ecophysiology

We aim

- to start by depicting habitats and plants physiognomically,
- to deduce problems from such observations in the field as are suited for physiological, biochemical and biophysical and perhaps even molecular experimentation in the laboratory, and
- to return from the laboratory to the field with increasingly sophisticated technologies for measurements and analyses applicable to field conditions.

With this we follow a great tradition, which was begun early in the last century, and which we may retrace from Mägdefrau's *History of Botany* (1992) as follows.

The title of one of the best-known essays (1806) by Alexander von Humboldt is *Ideas for a **Physiognomy** of Plants ("Ideen zu einer Physiognomik der Gewächse")*. He realized that the physiognomy of vegetation is determined by environmental conditions and that the distribution of plants depends on the climate and, thus he became the founder of **plant geography** (von Humboldt ed 1989). The selective pressure exerted by variation in environmental factors then also became the most essential aspect for explanation of **natural selection** in Charles Darwin's theory of evolution (1859). However, it was Ernst Haeckel who coined the term **ecology** in 1866. Stephen J. Gould (1977), the sharp American essayist in phylogeny, caricatured it as follows:

"Ernst Haeckel, the great popularizer of evolutionary theory in Germany, loved to coin words. The vast majority of his creations died with him half a century ago, but among the survivors are 'ontogeny', 'phylogeny', and 'ecology'. The last is now facing an opposite fate – loss of meaning by extension and vastly inflated currency. Common usage now threatens to make 'ecology' a label for anything good that happens far from cities or anything that does not have synthetic chemicals in it".

However, Ernst Haeckel's own original definition of ecology was already wide and may, in fact, encompass much of the current application of the term, since he wrote that ecology is "... the entire science of the relations of

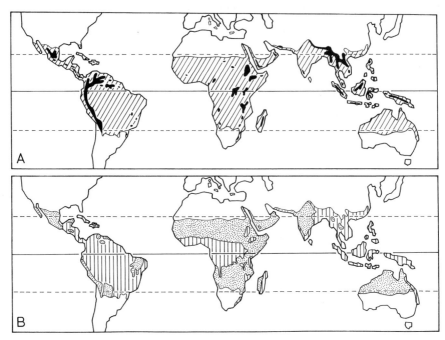

Fig. 1.1A, B. The hydrothermal partitioning of the tropics after Lauer (1975). **A** The warm tropics (*hatched*) and the cold tropics (*black*). **B** The wet tropics (*hatched*) and the dry tropics (*dotted*)

the organism to its surrounding environment, comprising in a broader sense all conditions of its existence".[1]

Andreas Franz Wilhelm Schimper (1856–1901) founded *Plant Geography on an Ecological Basis* with this title of his famous text published 1898, and he coined the term tropical rainforest. Simon Schwendener (1829–1919) suggested that the **relations between the environment and the morphological traits of plants** are best studied in areas providing extreme conditions. His advice has found many followers up to date. It was, however, Ernst Stahl (1848–1919) who introduced experimentation to ecological research and thus founded **physiological ecology.** He also discovered the role of stomata in transpiration and photosynthesis. Among the late scientists of our own century Arthur George Tansley (1871–1955), Otto Stocker (1888–1979), Arthur Pisek (1894–1975), Heinrich Walter (1898–1989), Bruno Huber (1899–1969) and Michael Evenari (1904–1989) all stimulated the development of physiological ecology.

---

[1] "... die gesamte Wissenschaft von den Beziehungen des Organismus zur umgebenden Außenwelt, wohin wir im weiteren Sinne alle Existenzbedingungen rechnen können."

1.2
The Tropics

Applying the approach of physiological ecology to understanding ecological functions and relationships in the **tropics** requires that we first must define what we mean by tropics.

Volkmar Vareschi (1980) has listed several possible definitions of the tropics, which partially overlap but give different emphasis:

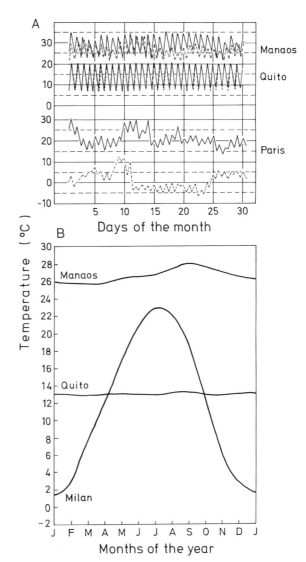

Fig. 1.2A, B. Diurnal (**A**) and annual temperature cycles (**B**) at a station in the warm tropics (Manaos at the Amazonas, Brazil) and in the cold tropics (Quito at 2660 m a.s.l. in the Andes, Ecuador) and at stations in the temperate zone (Paris and Milan, respectively). The diurnal cycles in **A** are given for the warmest months as indicated by *solid lines* (July, September and July for Manaos, Quito and Paris) and the coldest months are depicted by *dotted lines* (June, November and January for Manaos, Quito and Paris). (After Lauer 1975)

- geodetically the tropics are limited by the lines of latitude 23°27' north and south of the equator, i.e. the Tropic of Cancer and the Tropic of Capricorn respectively;
- climatologically the tropics are the zones of equal day and night length; they are not basically characterized by high temperature and moisture; depending on altitude, **warm and cold tropics** can be distinguished, and depending on the precipitation regimes in the region between the equator and the two lines of latitude at 23°27' north and south, **wet and dry tropics** can be distinguished (Lauer 1975; Fig. **1.1**).
- phyto-geographically the tropics are indicated by the distribution of palms;

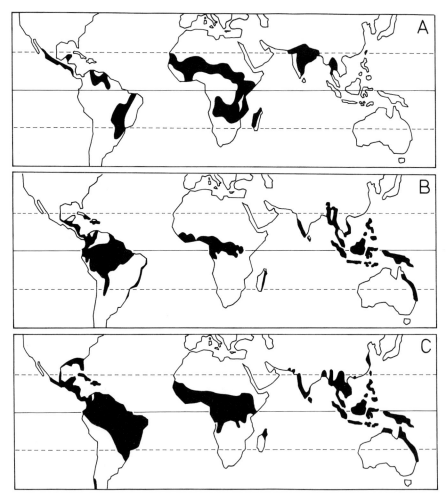

Fig. 1.3A–C. Global distribution of savannas (**A**) and tropical rainforest (**B**) and optimum carbon fixation (**C**). (After Vareschi 1980, **A, C** with permission of R. Ulmer; Walter and Breckle 1984, **B** with permission of S.-W. Breckle and G. Fischer-Verlag)

- eco-geographically the tropics are the zone in which the climatic effects of day-night cycles are far more important than those of seasonal cycles; day-night cycles of temperature are much larger than in the temperate zone in both the coldest and the warmest months, but annual cycles of mean monthly temperature are almost absent in the tropics (Fig. **1.2**);
- botanically the tropics are contained in a well-separated pantropical floristic province;
- in terms of the biology of productivity, the tropics are the zones of optimal carbon fixation and photosynthetic capacity with $>600 \, g \, m^{-2} \, a^{-1}$, which globally corresponds well to the occurrence of tropical rainforest in a broader sense (Fig. **1.3**).

1.3
Tropical Forests and Savannas: Their Emotional, Commercial, Ecological and Scientific Importance

Forests and savannas are the ecosystems which cover the largest areas in the tropics (Fig. **1.3**). Notwithstanding the wide public concern about destruction and decline of tropical forests, it appears that it is deeply engraved in the nature of man to be frightened by the unknown in the darkness of dense and repellant forests. There were incidents of natural catastrophes in which falling trees and landslides with forests threatened man, and after which a call arose even to remove the forest altogether. The contrast in our emotional reactions towards forest and savanna, respectively, is vividly expressed by Alexander von Humboldt (1982) in his *Journey to South America:*

"If one has spent many months in the dense forests along the Orinoco, if one got used there to seeing the stars only near the zenith like looking upwards from the bottom of a well as soon as one leaves the bed of the river, then wandering over the steppes[2] has something pleasant and attractive in it. The new pictures which one perceives give deep impression; like the Llanero one enjoys the feeling to be able to look around so well. However, this comfort is not of long duration. If, wandering for eight to ten days, one gets used to the games of mirages and the brilliant green of *Mauritia* bushes, which appear mile after mile, one feels in oneself the demand for more variable impressions, one longs for the sight of the huge trees of the tropics ..."[3]

---

[2] Note that A. von Humboldt did not yet distinguish between "steppes" and "savannas", which we now consider as the grasslands of temperate and tropical zones respectively (Vareschi 1980; Walter and Breckle 1984).

[3] "Hat man mehrere Monate in den dichten Wäldern am Orinoko zugebracht, hat man sich dort daran gewöhnt, daß man, sobald man vom Strome abgeht, die Sterne nur in der Nähe des Zenit und wie aus einem Brunnen heraus sehen kann, so hat eine Wanderung über die Steppen etwas Angenehmes, Anziehendes. Die neuen Bilder, die man aufnimmt, machen großen Eindruck, wie dem Llanero ist einem ganz wohl, 'daß man so gut um sich sehen kann'. Aber dieses Behagen ist nicht von langer Dauer. Ist man nach acht- oder zehntägigem Marsch gewöhnt an das Spiel der Luftspiegelung und an das glänzende Grün der Mauritiabüsche, die von Meile zu Meile zum Vorschein kommen, so fühlt man in sich das Bedürfnis mannigfaltigerer Eindrücke, man sehnt sich nach dem Anblick der gewaltigen Bäume der Tropen ...".

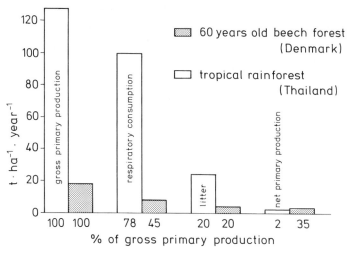

Fig. 1.4. Comparison of gross and net productivity of a tropical rainforest in Thailand and a 60-year-old beech forest in Denmark. (After data from Larcher 1980)

Commercially it is important to remember that most tropical countries have to sustain increasingly large populations, savannas serving agriculture and forests providing resources, which are, however, renewable only to a limited extent. Although original CO_2 sequestration and hence gross primary productivity of tropical forests is high, due to high rates of respiration and the rapid degradation of litter in the tropical forests net CO_2 uptake and net productivity is much reduced. It may even be lower than in a beech forest of the temperate zone (Fig. **1.4**).

Nevertheless, it is possible to reconcile utilization and preservation of tropical ecosystems if human activities are directed in the right way, as vividly summarized by Whitmore (1990). For example, there are two types of shifting agriculture (slash-and-burn agriculture; Fig. **1.5**). One of them is destructive and unsustainable. It is an invasive system, where fields are used until they are exhausted even for regrowth of secondary forest. The other one is a cyclic system, where clearings are used for cultivation for 1–2 years and then left to recover so that they can be reused in due course during the cycle, without the need for further clearing of forest. This is a sustainable mode of shifting agriculture. Figure **1.6** shows the recovery of above-ground biomass of a wet tropical forest after a slash and burn activity. Clearly, full recovery to a level comparable to mature original forest may take 100 to 200 years if the magnitude of disturbance has not been too large. However, if short-term cultivation after slash-and-burn action is followed by fallow periods of 20 years or longer, recovery mechanisms of forest ecosystems remain intact, and long-term cyclic utilization under low population pressure remains possible. On the other hand, large-scale deforestation proves to be irreversible (Medina 1991).

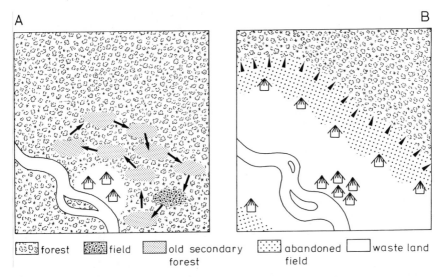

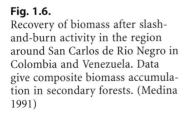

Fig. 1.5A, B. Two types of shifting agriculture. **A** Sustainable cyclic system. **B** Destructive and unsustainable invasive system. (After Whitmore 1990)

Fig. 1.6.
Recovery of biomass after slash-and-burn activity in the region around San Carlos de Rio Negro in Colombia and Venezuela. Data give composite biomass accumulation in secondary forests. (Medina 1991)

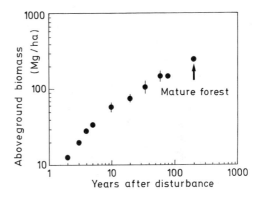

Scientifically sustainable schemes of silviculture have also been developed for certain tropical areas. They should be extended and enforced, including appropriate methods of logging and timber removal (Whitmore 1990). The tremendous difference in ecological quality between a more or less undisturbed woodland and a timber plantation becomes immediately clear at a glance (Fig. 1.7). However, afforestation and timber plantations can be established on degraded sites and will reduce the pressure on good natural forest. Minor forest products and agroforestry with intercropping of trees and foodcrops need to be seriously considered.

The ecological and scientific importance of savannas and forests in the tropics will be major topics of this book. There is a large diversity of unique

Fig. 1.7A, B. *Acacia* woodland (**A**) and *Eucalyptus* plantations (**B**) in the Rift Valley, Ethiopia

grassland and forest ecotypes, which require physiognomic classification and understanding at the ecophysiological level. This is a prerequisite for maintaining or, perhaps more pessimistically, for reattaining an equilibrium between commercial necessities including agriculture and forestry and the preservation of unique natural environments. It must be noted though, that no measures, even the most sophisticated schemes of sustainable land use based on scientific understanding of tropical biotopes and their components will be effective if man does not succeed in mastering the major dominating problem which is his own unlimited reproduction. Particularly in the tropics, increasing human population is the one major factor continuing to put pressure on the environment.

1.4
The Destruction of Tropical Forest

Of particularly wide public concern is the ongoing destruction of tropical rainforest. Estimations of the current destruction vary to some extent due to the application of different definitions of what actually is meant by tropical rainforest. This applies both to the question of what are the tropics (see above, Sect. **1.1**), and to the question of what is rainforest. (The latter is discussed in Sect. **3.1**.) Thus, in analyzing the problem, one may find oneself confronted with different quantitative estimates.

For example, by taking evergreen and semi-evergreen forests with

- no less than 100 mm rain in any month during 2 out of 3 years,
- an average yearly temperature of 24 °C without any occurrence of frost, and
- an altitude of <1300 m above sea level, excepting Amazonia with <1800 m and SE-Asia with <750 m,

and combining 13 critical and 3 very critical regions (as described in Fig. **1.8**), respectively, Myers (1988) arrives at the data given in Fig. **1.8**, or globally at the following numbers:

- forest preserved in 1980: 10×10^6 km^2, covering 6–7 % of the total land surface of the earth;
- disturbed: 0.1×10^6 km^2 per year;
- destroyed: 0.08 to 0.09×10^6 km^2 per year.

Using somewhat different definitions of tropical forests, Jacobs (1988) arrives at the following figures for the annual destruction:

- tropical rainforest in a narrow sense: 0.15×10^6 km^2 per year;
- wet tropical forest: 0.24×10^6 km^2 per year.

The major remaining areas covered with wet tropical forest are in the Zaire basin, in West Brazil and Amazonia, in the Guayana highlands and in New Guinea.

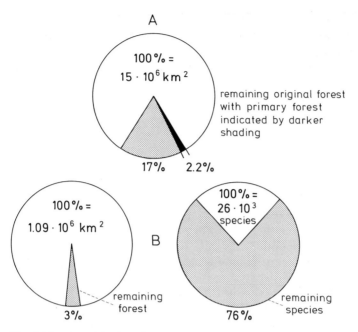

Fig. 1.8A, B. Original and remaining forest in 13 critical regions: (A Madagascar, Atlantic coast of Brazil, West Ecuador, Columbian Chocó, West Amazonian Highlands, Rondonia/Acre in Brasilian Amazonia, montane forests in Tanzania/Kenya, Eastern Himalaya, Sinharaja Forest in Sri Lanca, Malaysian peninsula, NW-Borneo, Philippines and New-Caledonia, and in three particularly critical regions: **B** Madagascar, Atlantic coast of Brazil and West Ecuador. (After data of Myers 1988)

Some important global problems relate to the destruction of tropical forest with regard to

- the CO_2 budget of the atmosphere,
- the water balance,
- the nutrient balance, and
- biodiversity,

the first two and the last of which are also causing considerable public anxiety.

Scientifically, the effects on CO_2 budgets remain a subject of debate because it is not clear whether alternative CO_2 fixation processes in terrestrial ecosystems will offset or even overcompensate reductions due to loss of forest. Secondary vegetation may prove to be an increasingly strong CO_2 sink, and increasing CO_2 in the atmosphere may be coupled to higher ecosystem productivity (Medina 1991; Plant, Cell and Environment 1991).

The water balance of large areas may be severely impaired by deforestation. For equatorial forests in Amazonia it has been shown by stable-isotope techniques (see Sect. **2.5**) that 50 % of total incoming rainfall was lost again by evapotranspiration from the forest. Thus, deforestation not only increases total runoff of water but also disturbs recirculation, as observed in the Amazon basin, leading to lower total rainfall and more pronounced seasonality (Medina 1991).

These observations also have implications for nutrient supply: due to the rapid turnover of nutrients in tropical forests (see Sect. **3.5**) and problems of erosion, deforestation causes major destruction of soil systems and affects nutrient budgets.

Tropical humid forests are known to be the most diverse ecosystems in the world (see Sect. **3.3.1**). They are thought to support more than 50 % of all plant and animal species. Deforestation leads to loss of diversity, which has not been fully assessed by census to date.

In conclusion, we are not able to predict the actual nature of all the changes that may result from the more or less complete destruction of these forests (Whitmore 1990). The theory of deterministic chaos (Sect. **3.3.1**) suggests that long-term predictions about the behaviour of complex systems with feedback relations showing non-linear behaviour are intrinsically impossible (Hastings et al. 1993; Schuster 1984). However, it is equally clear that if we do not succeed in preserving these forests, we shall lose one of our greatest treasures.

References

Darwin C (1859) On the origin of species by means of natural selection. John Murray, London

Gould SJ (1977) Ever since Darwin. Reflections in natural history. Penguin Book Harmondsworth

Hastings A, Hom CL, Ellner S, Turchin P, Godfray HCJ (1993) Chaos in ecology: is mother nature a strange attractor? Annu Rev Ecol System 24:1–33

Jacobs M (1988) The tropical rain forest. Springer, Berlin Heidelberg New York

Larcher W (1980) Ökologie der Pflanzen auf physiologischer Grundlage. Ulmer, Stuttgart

Lauer W (1975) Vom Wesen der Tropen. Klimaökologische Studien zum Inhalt und zur Abgrenzung eines irdischen Landschaftsgürtels. Akad Wiss Lit Mainz Abh Math Naturwiss Kl 1975/3:5–52

Mägdefrau K (1992) Geschichte der Botanik. Leben und Leistung großer Forscher. 2. Aufl. G Fischer, Stuttgart

Medina E (1991) Deforestation in the tropics. Evaluation of experiences in the Amazon basin focussing on atmosphere-forest interactions. In: Mooney HA et al. (eds) Ecosystem experiments. John Wiley, New York, pp 23–43

Myers N (1988) Tropical forests and the botanists' community. In: Greuter W, Zimmer B (eds) Proc XIV Int Botanical Congr. Koeltz, Königstein, pp 291–300

Plant cell and environment (1991) Special issue, elevated CO_2-levels, vol 14, no 8. Blackwell Science Ltd., Oxford

Schuster HG (1984) Deterministic chaos. Physik Verlag, Weinheim

Vareschi V (1980) Vegetationsökologie der Tropen. Ulmer, Stuttgart

von Humboldt A (1982) Südamerikanische Reise. 1808, quoted after the edition of Greno. Verlagsgesellschaft mbH, Nördlingen

von Humboldt A (1989) Schriften zur Geographie der Pflanzen. In: Alexander von Humboldt Studienausgabe Band I, Beck H (ed) Wissenschaftliche Buchgesellschaft, Darmstadt

Walter H, Breckle S-W (1984) Ökologie der Erde, vol 2. Spezielle Ökologie der tropischen und subtropischen Zonen. G Fischer, Stuttgart

Whitmore TC (1990) An introduction to tropical rain forests. Oxford University Press, Oxford

CHAPTER 2

Large-Scale Sensing and Diagnosis in Relation to the Tropical Environment

2.1
Approaches

In the discussion of "global change" it has become increasingly important to develop means allowing large-scale conclusions about the conditions and the behaviour of ecosystems or biomes. Indeed, techniques for examination and detailed structural analysis of the surface of our globe are continuously advanced, allowing the integration of observations in space and time. This is, of course, applicable throughout the globe. However, it is particularly relevant for the tropical environment, with extended tracts of ecosystems like savannas and forests (see Sect. 1.3), which are often difficult to penetrate on the ground.

Among the techniques applicable to the study of large-scale effects of plant life we must distinguish between those, where the analytical method itself directly covers wide spaces and areas, such as remote sensing (Sect. 2.3), and those, where it is the method of sampling, which can be developed to cover extended areas, and thus, indirectly affords comprehension at large-scale levels. In the former case only a smaller choice of analytical techniques mainly based on the use of radiation will be available, while in the latter case almost any analytical method may prove useful depending on availability of suitable samples. Clearly, there are many approaches for large-scale sensing and diagnosis of the environment, some of which shall be discussed below.

2.2
Climatic Relations and Vegetation Modeling

2.2.1
The *Klimadiagramm*

As the primary producers of biomass, plants determine the physiognomy and character of large ecological units and provide the basis for all other life. This has contributed greatly to the practice of using patterns of vegetation for a coarse global division of the geo-biosphere in ecological terms. On the other hand, Walter and Breckle (1983) argue that the large naturally occurring communities of flora and fauna, or "**biomes**", are best delineated on the basis of the climatic conditions under which they occur, since "the climate ... is the only primary factor, which affects the other factors like the soil and the vegetation and to a lesser degree also the fauna, but which in turn is influenced by them only to a limited extent in the range of micro-climate". In global terms a coarse division of geographical regions is given by the large climatic **zones**, from which Walter and Breckle (1983) then derived their term **zono-biomes.** Thus, we also obtain the large zones of vegetation in a **three-dimensional climatic gradation** (Ehrendorfer 1991):

- from the equator to the poles (temperature gradient),
- from the oceans to the continents (oceanity or continentality according to the degree of annual balance of temperature), and
- in the altitudinal zones in high mountains.

This possibility of separating large-scale vegetation units according to the climate leads to the practical question of whether one can make predictions on plant distribution in extended areas based on simple models.

One simple technique, which has proven extraordinarily successful, is the **Klimadiagramm** of Walter (1973). In his autobiography Walter (1982) describes vividly how the idea developed when he was first confronted with the problem of a large-scale interpretation of vegetation in Anatolia (Turkey) in 1954. The *Klimadiagramm* (Box **2.1**) essentially uses simple data which are readily available from all weather stations, i.e. mean monthly temperatures and precipitation. They are plotted according to a precise scheme of scaling on the ordinate versus months on the abscissa: **humid periods** are indicated by areas on the graph, where the temperature curve is below the curve of precipitation; **arid periods** are delineated by the precipitation curve being lower than that of temperature. According to a precisely defined scheme, additional information can be built into the *Klimadiagramm*, so that depending on data availability each diagram may give a complete description of the climate at a given station.

For a large-scale evaluation of a certain area, or even an entire continent, one requires the diagrams of many stations covering the respective area. One can integrate them in a geographic information system, e.g. in the simplest way paste them on a map to obtain a good survey of the climatic structure of the area. Areas with humid, arid or seasonal climates are readily separated. An example is shown for the predominantly tropical continent of Africa in Fig. **2.1**. The humid belt along the equator is clearly separated from the more semi-arid and arid regions. The power of the approach can be seen by comparing the distribution of rainforest and savanna on the African continent (Fig. **1.3**) with the more humid and more seasonal regions, respectively (Fig. **2.1**).

Although we will have to draw attention to certain limitations of the *Klimadiagramm* technique later (Sect. **3.2**), since arid and humid climates are strictly defined by precipitation and evaporation and not by precipitation and temperature as in the *Klimadiagramm*, we will make repeated use of the *Klimadiagramm* in this book.

Box 2.1

Klimadiagramm after Walter (1973)

- The months of the year are plotted on the abscissa.

- The mean monthly temperatures and precipitation are plotted on the ordinate, so that
  - at mean monthly precipitation between 0 and 100 mm one unit of scale corresponding to 10 °C gives 20 mm precipitation or the ratio of scalation is 1 °C : 2 mm precipitation;
  - at mean monthly precipitation above 100 mm the precipitation scale is reduced to 1/10, and the ratio is 1 °C : 20 mm precipitation.

- Humid periods are indicated by precipitation curves above temperature curves; they are marked by *vertical hatching* up to 100 mm and by *black colour* above 100 mm precipitation.

- Arid periods are indicated by precipitation curves below temperature curves, they are marked by *dotting*.

- According to a well-defined scheme, other details may be added to the *Klimadiagramm* as indicated in the examples given below.

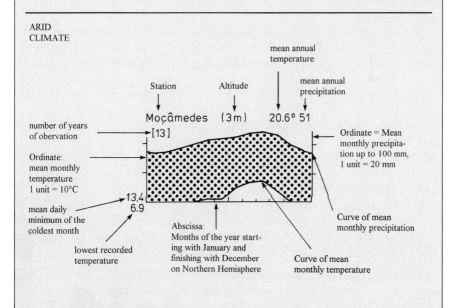

ARID CLIMATE

mean annual temperature

Station   Altitude   mean annual precipitation

Moçâmedes (3 m)   20.6° 51

number of years of obervation → [13]

Ordinate: mean monthly temperature 1 unit = 10°C

Ordinate = Mean monthly precipitation up to 100 mm, 1 unit = 20 mm

13.4
6.9
mean daily minimum of the coldest month

Abscissa: Months of the year starting with January and finishing with December on Northern Hemisphere

lowest recorded temperature

Curve of mean monthly precipitation

Curve of mean monthly temperature

Box 2.1 (Continued)

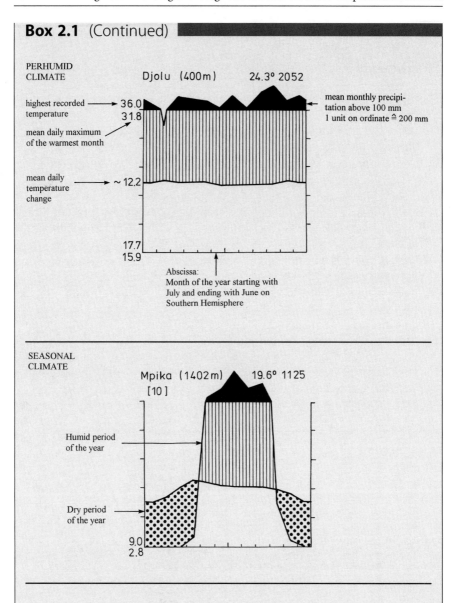

PERHUMID
CLIMATE

Djolu (400m)   24.3° 2052

highest recorded ——► 36.0
temperature   31.8

mean monthly precipi-
tation above 100 mm
1 unit on ordinate ≙ 200 mm

mean daily maximum
of the warmest month

mean daily   ~ 12.2
temperature
change

17.7
15.9

Abscissa:
Month of the year starting with
July and ending with June on
Southern Hemisphere

SEASONAL
CLIMATE

Mpika (1402m)   19.6° 1125
[10]

Humid period
of the year

Dry period
of the year

9.0
2.8

Examples presented are for tropical stations in Africa with an arid, a per-
humid and a seasonal climate: Moçâmedes at the Atlantic coast of
Angola (15°05' S, 12°09' E), Djolu, Congo (00°38' N, 22°37' E) and Mpika,
Muchinga Mountains, Rhodesia (11°52' S, 31°26' E). For further details see
Walter and Lieth (1967) and Walter (1973).

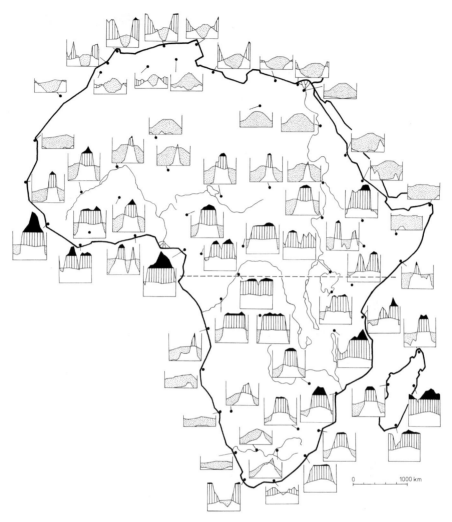

Fig. 2.1. *Klimadiagramm* map of the African continent. Arid periods are *dotted*, humid periods are *hatched* or *black*. Arid, humid and seasonal regions are readily differentiated. (Walter and Breckle 1984; with kind permission of S.-W. Breckle and G. Fischer-Verlag)

2.2.2
Vegetation Modeling Based on Irradiance and Water Budgets

Climatic conditions like **irradiance** and **temperature** affect the **water-vapour pressure deficit of the atmosphere**, and thus determine **evapotranspiration** or the loss of water vapour of the vegetation to the atmosphere. Irradiance and water availability in turn are modulated by other climatic factors. Thus, it is possible to make model calculations of evapotranspiration from climatic data using basic plant-physiological principles of transpiratory water loss from leaves driven by the leaf/air water-vapour pressure-difference. Furthermore, there are close links between evapotranspiration and **leaf area index** (**LAI**), which is the total leaf area related to a unit of ground surface (see Sect. **3.2.2**). The LAI is characteristically related to the

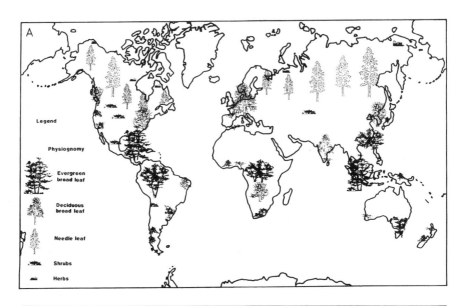

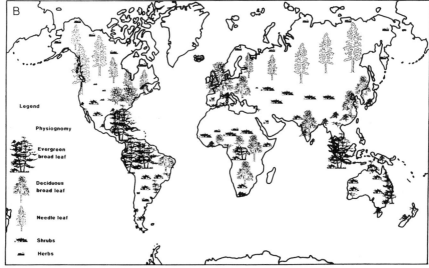

Fig. 2.2A, B. **A** Global prediction of physiognomic vegetation types on the basis of eco-physiological models; compared with **B** actual observations. (Woodward 1987; with kind permission of the author and Cambridge University Press)

physiognomy of plants and vegetation. Therefore, from LAI one may then obtain ecophysiological models of vegetation as dominated respectively by broad-leaved trees, shrubs and herbs (Woodward 1987). Large-scale presentations of the models' results may then be compared with real observations (Fig. **2.2**).

2.3
Remote Sensing Using Radiation

Remote sensing of the biosphere is based on the analysis of electromagnetic radiation (Hobbs and Mooney 1990). We may distinguish between measurements of reflection or absorption of radiation and fluorescence. Radiation is also used in gas analysis, but this is a somewhat different aspect (see Sect. **2.4**).

2.3.1
Reflection and Absorption

The analysis of **reflection, absorption and transmission of radiation** by individual leaves, plants or by the vegetation canopy has become an important method in ecophysiology. The principle relations are shown in Fig. **2.3**. The range of photosynthetically active radiation (PAR: 400–700 mm) is largely identical to that of the visible light. Here **radiation absorption** is dominating. Chlorophyll has an absorption minimum in the green range of the spectrum (550 nm), and this is identified by reduced absorption, and increased reflection and transmission (Fig. **2.3**). In the infrared range of the spectrum (above 800 nm) **radiation reflection and transmission** are dominating. At very high wavelengths absorption increases again, although, this is not so relevant as solar emission contains little radiation above 2000 nm.

Fig. 2.3.
Relationship between reflection, absorption and transmission of radiation by a green leaf at varied wavelengths. *UV* Ultraviolet; *PAR* photosynthetically active radiation; *IR* infrared. (After Gates 1965; from Nobel 1983: Biophysical Plant Physiology and Ecology, Copyright 1983 by W.H. Freeman and Company; used with permission)

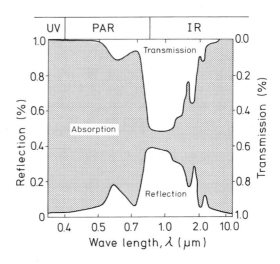

The contrast between absorption of the radiation in the visible range and reflection in the infrared range of the spectrum by green plants can be used to develop a dimensionless **vegetation index** Q related to reflection between 580 and 680 nm (R_{580}^{680}) and between 725 and 1100 nm (R_{725}^{1100}) respectively as follows:

$$Q = \frac{R_{725}^{1100} - R_{580}^{680}}{R_{725}^{1100} + R_{580}^{680}} \qquad (2.1)$$

It results from this equation that at very low reflection between 580 and 680 nm and very high reflection between 725 and 1100 nm, vegetation is dense and Q tends towards +1. In contrast, at very high reflection between 580 and 680 nm and very low reflection between 725 and 1100 nm Q tends towards -1, indicting sparse vegetation (Running 1990). The two values of R, R_{580}^{680} and R_{725}^{1100}, can be measured from aeroplanes or meteorological satellites equipped with two sensors for the respective range of wavelengths. The results can be depicted on maps using false colours, which provide informative images at the global level. Formations with particularly dense vegetation (e.g. the tropical forests and the extended forest regions of the northern hemisphere) are readily distinguished from poorer areas like deserts, steppes and savannas (Malingreau and Tucker 1987).

The disadvantage in the analysis of radiation reflection is that clear skies are needed. An alternative is the use of **micro-wave radiometry**, since micro-waves are not absorbed by clouds. They give information on the heat-radiation of surfaces, which is also much influenced by vegetation.

In principle remote sensing by detection of radiation allows analysis of many vegetation parameters on a large scale, e.g.

- the vegetation density on the land surface according to the leaf area index and the biomass;
- the density of plankton in the oceans;
- vegetation types and the structure and dynamics in ecosystems;
- the productivity of agricultural and natural ecosystems;
- phenological cycles of growth;
- gross biochemical composition of vegetation utilizing the light absorption of quantitatively dominant organic compounds (e.g. sugars, cellulose, starch, lignin, protein) measured with spectrometers of particularly high resolution (Wessman 1990);
- soil moisture;
- hydrology;
- changes by deforestation, fire, storm, erosion, pest infestations and other catastrophic events.

In many of these examples analysis of temporal variations is essential, integrated by repeated observation. Another important aspect is the need for "**calibration**", which may allow a more detailed interpretation of the radiation signals obtained in remote sensing. Due to the complex structure and dynamics of ecosystems, significant efforts are required for calibration, involving measurements on a hierarchy of levels with different resolution of area, e.g. on the ground, on measuring towers, in aeroplanes and helicopters and in satellites (Sellers et al. 1990). With better calibration, interpretation of finer detail and more sophisticated resolution will be possible from remote-sensing data. However, calibration or "ground truthing" itself is a great problem, especially in inaccessible tropical areas.

2.3.2
Fluorescence

The analysis of reflection and absorption provides information on a given state of vegetation and, if followed in time, also about its dynamics. Although this type of analysis has already been used successfully to predict harvests, it does not provide a real picture of the **physiological state and vitality of vegetation.** This contrasts with analysis of fluorescence. The fluorescence from chlorophyll is not only directly related to the concentration of chlorophyll but is also inversely related to the **efficiency of photosynthesis.** This will be explained in more detail later in relation to photosynthetic light use in tropical environments (Sect. **3.6.4**).

Remote fluorescence excitation is possible using powerful lasers. A laser-induced fluorescence spectrum is given in Fig. **2.4**. The fluorescence maxima at 690 nm and 740 nm are particularly variable in response to stress as shown in Fig. **2.5** for some conditions studied in the laboratory. As plants senesce fluorescence decreases due to the degradation of chlorophyll (Fig. **2.5 A**). Conversely, when the photosynthetic process is impaired, e.g. by K^+ deficiency, drought stress or herbicide action (Fig. **2.5 B–C**), fluorescence increases (Chapelle et al. 1984a). For a more sensitive analysis it has also been suggested that the ratio of fluorescence at the peak of 690 nm and in the far-red region at 730 or 740 nm should be used (Hák et al. 1990; Lichtenthaler et al. 1990).

Thus, when used in remote sensing, fluorescence analysis allows large-scale diagnosis of stress effects by abiotic factors such as the availability of

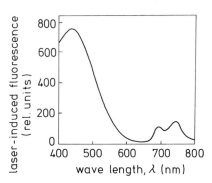

Fig. 2.4.
Laser-induced fluorescence spectrum of maize leaves (Chapelle et al. 1984b)

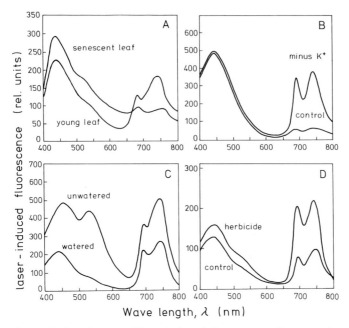

Fig. 2.5A–D. Changes of laser-induced fluorescence due to various kinds of stress. **A** Senescence. **B** Potassium deficiency. **C** Drought. **D** Action of the herbicide DCMU (dichlorophenyl-dimethyl-urea) inhibiting photosynthesis. **A, C, D** soybean; **B** maize. (Chapelle et al. 1984a)

growth resources (e.g. water, mineral nutrients, photosynthetically active radiation etc.) or environmental pollutants and biotic factors, such as pests and pathogens. The resolution of analyses of laser-induced fluorescence of 685 nm during flights in meteorological aeroplanes flying at a height of 150 m and a nominal flight-speed of $100\,m\,s^{-1}$ is between 10 and 80 m. An example is given in Fig. **2.6** for a flight path of 6 km, resolving green and brown fields and forests.

2.4
Gas Analysis

Another means of analyzing the effects of plant life across a range of scales is that of **infrared gas-analysis (IRGA)**. It is of great importance because the method can measure the **gas exchange of plants**, particularly respiratory and photosynthetic CO_2 exchange (but not O_2!) and also transpirational loss of water vapour. Many molecules which play a role as **environmental pollutants**, such as sulfur dioxide, nitrogen oxides, ammonia and carbon monoxide can also be analyzed. Therefore, the IRGA is an important technique in ecophysiological studies encompassing photosynthesis as well as environmental control.

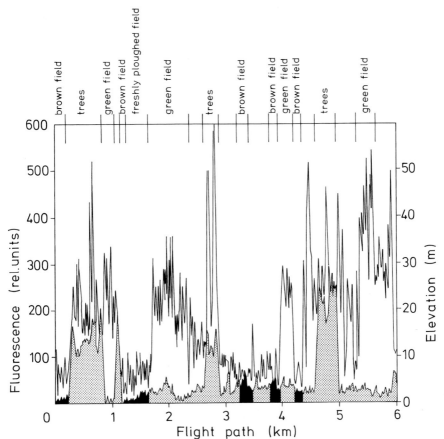

Fig. 2.6. Profile of laser-induced fluorescence emission (wavelength of fluorescence 685 nm) along a flight path of 6 km above forests and fields. The *bright profile* gives fluorescence. The *dark profile* indicates terrain elevation and at the same time the different parts of the landscape, i.e. green fields, brown and freshly ploughed fields and trees. (After Hoge et al. 1983)

An IRGA measures the specific absorption of **infrared radiation** with a characteristic absorption spectrum for a particular gas. It is caused by an uneven distribution of electrical charge in the gas molecules, i.e. a dipole moment, which can be excited by infrared radiation. This is present in gas molecules which are composed of two or more different atoms (Box **2.2**). Moreover, gas molecules which are built up of more than two identical atoms also develop a dipole moment due to the oscillations of atoms against each other. For example in ozone, composed of three oxygen atoms (O_3), two of the atoms are always closer together and the distance to the third atom is larger (induced dipole moment). Conversely, gases with two identical atoms are not sensitive to infrared radiation, because even during oscillation of atoms the charge remains evenly distributed. (A list of important infrared active and non-active gas molecules is given in Box **2.2**.)

Box 2.2

Infrared active and non-active gases

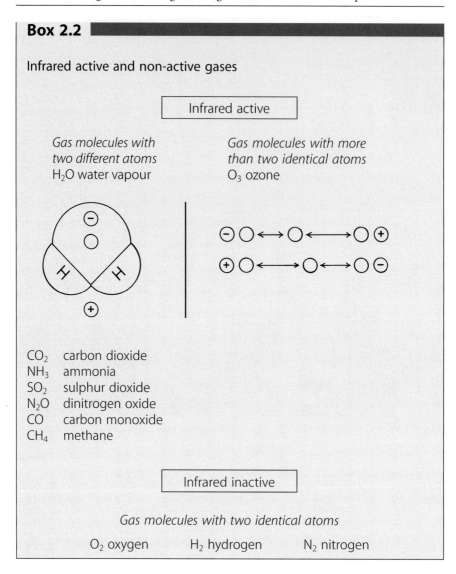

| Infrared active |
| Gas molecules with two different atoms H_2O water vapour | Gas molecules with more than two identical atoms O_3 ozone |

CO_2 carbon dioxide
NH_3 ammonia
SO_2 sulphur dioxide
N_2O dinitrogen oxide
CO carbon monoxide
CH_4 methane

| Infrared inactive |

Gas molecules with two identical atoms

O_2 oxygen H_2 hydrogen N_2 nitrogen

For the same reason which allows their detection by IRGA, the infrared-active gases also contribute to the green-house effect, keeping the surface of the globe warm by absorbing infrared heat radiation which would otherwise radiate out into space. The most important gas in this respect is water vapour, without which the earth would be unbearably cold. The recent increase of other atmospheric gases like CO_2 and methane (CH_4) in the last hundred years is considered to threaten global temperature balance.

Fig. 2.7 shows how photosynthetic or respiratory gas exchange of individual leaves or plants is measured in gas-exchange cuvettes. Studies at larger scales can be made by taking samples during meteorological flights with aeroplanes or balloons, which are later analyzed by IRGA-techniques. Such

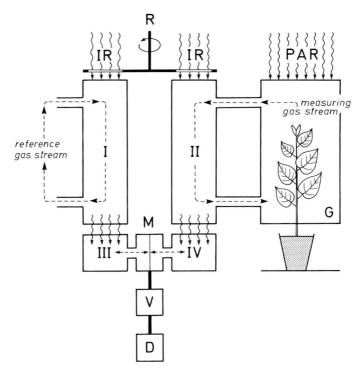

Fig. 2.7. The principle of infrared gas analysis (IRGA): The gas analyzer consists of four gas-filled chambers *I–IV*. The gas to be analyzed is passed through chamber II. It is often a gas stream modified during photosynthesis by leaves in a gas-exchange cuvette (*G*) under photosynthetically active radiation (*PAR*). The reference gas is contained in chamber I. An infrared radiation (*IR*) incident on both chambers is absorbed in correlation to the gas concentrations in the chambers. The radiation transmitted by chambers I and II heats up the gas contained in chambers III and IV below them. Depending on the intensity of the transmitted radiation, which is influenced in chamber II by the concentration in the measured gas, the gas in chambers III and IV is heated up to a larger or smaller extent. This causes a pressure difference which moves the membrane of a membrane condensor (*M*). The IR beams are chopped by a rotating disc (*R*), which interrupts the radiation in both chambers in an equal rhythm. Thus, synchronized changes of capacity and potential are obtained, which are amplified and rectified (*V*) and given out on a recorder, into a data logger (*D*) or on a digital read-out unit. With the appropriate choice of IR wave lengths corresponding to the IR absorption spectrum of the gas species studied, different gases can be analyzed. Since chambers III and IV also must contain the gas species to be measured, analysis of each gas species requires a separate specific analyzer. Cross-sensitivities for other gases in some cases may also have to be considered

flights in the lower atmosphere are increasingly valuable in the investigation of large-scale ecological problems. An impressive example of the success of this approach is the study of gas exchange above the canopy of a tropical forest by measuring CO_2-concentrations. It has been demonstrated that high CO_2-concentrations build up above the canopy from respiration during the night and are decreased in photosynthesis during the day (Fig. **2.8**).

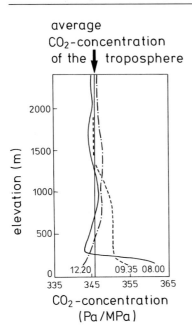

average
CO_2-concentration
of the ↓ troposphere

Fig. 2.8.
Vertical profiles of CO_2 concentrations in the atmosphere above a tropical forest based on the infrared analysis of gas samples taken at three different times during the morning. At 08.00 h CO_2 concentration is high due to nocturnal respiration of the plants; however, it gradually decreases in the course of the morning due to photosynthesis, and at 12.20 h it is clearly smaller than the average CO_2 concentration of the troposphere at a little more than 345 Pa/ MPa. (After Matson and Harriss 1988)

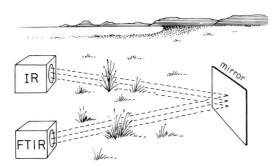

Fig. 2.9.
Infrared-radiation source (*IR*) and Fourier transform infrared spectrometer (*FTIR*), which together with telescopes and a mirror (or a network of mirrors) allow ground-based gas analysis over longer distances. (After Gosz et al. 1988)

Large areas can also be investigated by ground-based gas-analysis systems spanning a range from meters up to kilometers. Such an instrument, for example, is the fourier-transformed-infrared-spectrometer (FTIR), which is equipped with a 60 cm telescope and appropriate computer technology, and using mirrors can provide trace-gas analyses over distances of up to 1.5 km (Fig. **2.9**) (Gosz et al. 1988). Of course, this approach is difficult in denser and taller vegetation. New laser sensors, which operate not only in the infrared region of the spectrum but also in the visible and ultraviolet region, allow direct gas analyses from onboard aeroplanes (Matson and Harriss 1988; Matson and Vitousele 1990). Here, the troposphere at the surface of our planet is taken as kind of a gas-exchange cuvette to consider gas-exchange between the biosphere and the atmosphere. The example of Fig.

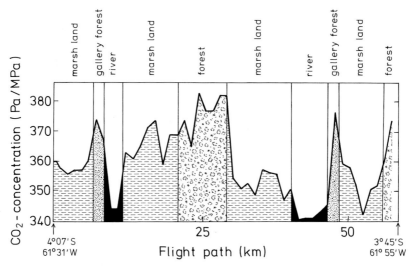

Fig. 2.10. Horizontal profile of CO_2 concentrations in the atmosphere above different ecosystems in the tropics based on measurements on board of a meteorological aeroplane along a flight path of a little more than 50 km between 08.30 and 08.43 h. The vegetation has increased the CO_2 concentration due to nocturnal respiration. The values above rivers are much lower and correspond to the average CO_2 concentration in the troposphere (see also Fig. 2.8). (After Matson and Harriss 1988)

2.10 shows the separation of different atmospheric CO_2 concentrations above various tropical ecosystems along a 50-km-long path of a flight in the morning.

2.5
Stable Isotope Analysis

Many elements in nature occur in the form of several **isotopes**, which for a given proton content in the nuclei have slight differences in mass due to varied neutron contents. In most elements one of the isotopes quantitatively predominates, while the others have a much lower abundance.

Isotope techniques initially became well known through **radioactive isotopes** serving as tracers to follow the path of certain elements in complex metabolic reaction sequences and transport systems. Alternatively, radiocarbon dating uses the naturally occurring carbon isotope ^{14}C, with a half-life of appr. 5600 years, and provides a means of dating over relatively recent geological time scales. More recently, however, the analysis of **stable isotopes** has provided increasingly valuable information on spatial and temporal variations in ecological and palaeohistorical terms (Rundel et al. 1988). Table 2.1 provides information on those stable isotopes which currently promise to provide the most important information in ecology.

Element	Isotope	Abundance (%)
Hydrogen	$_1$H	99.985
	^{2}H or D	0.015
Carbon	^{12}C	98.89
	^{13}C	1.11
Nitrogen	^{14}N	99.63
	^{15}N	0.37
Oxygen	^{16}O	99.759
	^{18}O	0.204
Sulphur	^{32}S	95.00
	^{34}S	4.22
Strontium	^{86}Sr	9.86
	^{87}Sr	7.02
	^{88}Sr	82.56

Table 2.1.
A small selection of stable isotopes which promise to be important tools in ecology. (Ehleringer and Rundel 1989)

The differences in the physical and chemical properties of the various isotopes of an element depend on the relative differences in mass and therefore they are small. The exception is hydrogen with very light nuclei where the deuterium provides much greater isotope effects. However, the analytical methods are extraordinarily precise. The isotopes are analyzed using **mass-spectrometry** according to their mass to charge ratio. The results of the analysis of a gaseous sample are not expressed as absolute isotope contents but related to an internationally accepted standard. Let x' be a rare and x the most frequent isotope of an element, where x' may be heavier or lighter than x, one can obtain the **isotope-ratio** δx' in o/oo as follows:

$$\delta x' = \left(\frac{(x'/x)_{\text{sample}}}{(x'/x)_{\text{standard}}} - 1 \right) \times 1000 \ . \tag{2.2}$$

Thus, in fact, one is operating with isotope-ratios, which provides the high precision in comparisons of different samples, expressed as a differential against the particular standard.

Isotope effects, are a measure of the behaviour of individual molecules containing different isotopes of an element and occur during diffusion or transport in various media as well as in enzymatic reactions. The differences in the kinetic properties and in the behaviour of molecules with different isotopic composition in thermodynamic equilibria form the basis for many applications of the stable-isotope technique, including

- geochemistry,
- hydrology,

- meteorology,
- paleoclimatology,
- biochemistry and physiology of metabolism,
- ecology and environmental research.

The technique originated from geochemistry. Its great importance in meteorology may be illustrated by considering **water.** The vapour pressure of water is proportional to the mass. Thus, heavy H_2O-molecules with the isotopes ^2H and ^{18}O instead of ^1H and ^{16}O need higher temperatures to evaporate and are discriminated against. However, at higher temperatures the absolute H_2O content in the gas-phase is higher. During rain the heavier molecules then precipitate more readily than the lighter ones, and hence, evaporation and precipitation provide a **climate effect** on isotope composition of rain (Ziegler 1989):

- according to **latitude;**
  the content of heavy isotopes ^2H and ^{18}O declines with increasing latitude following the temperature gradient;
- according to **altitude;**
  the content of ^2H and ^{18}O declines with increasing altitude;
- according to **seasons;**
  at higher latitudes ($>30°$) the rain contains more ^2H and ^{18}O in summer and less in winter;
- according to **continentality;**
  the ^2H and ^{18}O contents decrease with increasing distance from the coast;
- according to the **amount of rain;**
  the ^2H and ^{18}O contents decrease with increasing amount of rain falling.

This explains **hydrological isotope effects.** The isotope composition of groundwater, flowing surface-water, and recent precipitation is different. Thus, although no fractionation occurs during uptake by plants, organisms which take up such water are also distinguished by their own isotope content. Although there are additional discrimination processes when water evaporates from leaves (enriching leaf ^2H and ^{18}O), a further fractionation occurs in favour of the heavy isotopes during incorporation into organic material. By analysis of the most recently produced biomass, allowing for certain exchange reactions one can conclude from which direction the last rain falls or one can determine the geographical origin of plants and food items (Smith 1975).

These climate effects have led to applications in **palaeoclimatology.** From the isotope composition of fossil water in subterranean water reservoirs, which have formed in geological periods, one can draw conclusions on the climate at that time. Analysis of small gas inclusions in layers of ice in glaciers, or of water from the ice itself, allow similiar conclusions about the past. For example, in the ice of a glacier at 5670 m a.s.l. in the Cordillera de

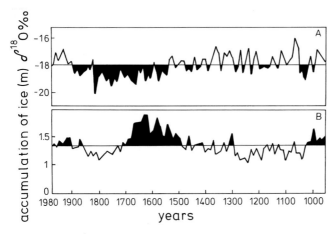

Fig. 2.11A, B. δ^{18}O-values (**A**) [see Eq. (2.2)] and accumulation of ice (**B**) for the past 1000 years in the glacier Quelcaya Cap, Cordillera de Carabaya, Peru (13° 56' S, 70° 50' W) related to values of the year 1980. (After Jones 1990; with kind permission by La Recherche)

Carabaya of the tropical Andes in Peru, relatively low levels of ^{18}O [more negative values of δ^{18}O, see Eq. (2.2)] indicate a warmer period between 1000 and 1500 A.D. This was correlated with lower accumulation of ice, although higher ^{18}O levels (less negative δ^{18}O), then represent a colder period from 1500 to about 1875 A.D. when ice-deposition was initially high (Fig. **2.11**). Further opportunities are provided by the incorporation of water into organic constituents of plants, and the analysis of gross remnants of plants in peat accumulations or from the study of isotope composition of annual rings in old trees and wood.

The most sophisticated applications of the stable-isotope technique in ecology result from **isotope effects of metabolism.** Several enzymes discriminate against substrate molecules constituted of different isotopes leading to different isotope composition of products. An important example is the fixation of CO_2 for photosynthetic assimilation with discrimination occurring against the heavier isotope ^{13}C as compared to ^{12}C. The enzyme ribulosebisphosphate carboxylase (RuBPC), which directly leads to the formation of the C_3-compound phosphoglyceric acid in the light (**C_3-photosynthesis**), discriminates against $^{13}CO_2$ more strongly than another carboxylating enzyme, phosphoenol-pyruvate carboxylase (PEPC), which forms the C_4-acid oxaloacetate and leads to malate/aspartate (**C_4-photosynthesis**, Box **7.2**). The latter enzyme (PEPC) also mediates dark fixation of CO_2 in plants with **crassulacean acid metabolism** (CAM, Box **3.11**), where malic acid provides nocturnal CO_2 storage. In CAM, during the day, CO_2 is remobilized from malate and refixed and assimilated via RuBPC. Depending on the environmental conditions CAM-plants can also fix CO_2 in the light directly via RuBPC.

Fig. 2.12.
Ranges of carbon-isotope ratios for
C_3-, C_4- and CAM species. Measurements from 285 grasses (Poaceae)
with 47 C_3 species and 238 C_4 species and from 513 CAM plants.
(Unpub. data sets of H. Ziegler)

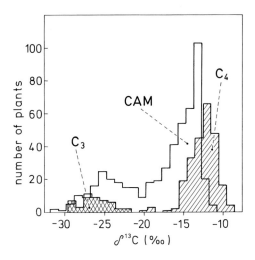

Due to the different extent of discrimination against $^{13}CO_2$ by the two
enzymes during primary CO_2-fixation by the C_3- and C_4-plants, and by the
variable expression of the two mechanisms in CAM-plants, **carbon-isotope
ratios** ($\delta^{13}C$ values) differ in the three types of plants. They are most negative (most depleted of ^{13}C) in C_3-plants, less negative in C_4-plants and intermediate in CAM-plants (Fig. **2.12**). Both C_4-photosynthesis and CAM are, in
different ways, adaptations to reduced supply of water and contribute to the
efficiency of plant water use. This means that in C_4- and CAM-plants, for
each CO_2-molecule fixed, a smaller number of H_2O-molecules is lost via
transpiration as compared to C_3-photosynthesis. Thus, in addition to the
^{13}C-content, the ^2H- and ^{18}O-content of the plants is also affected. Even
within the different modes of photosynthesis there are subtleties dependent
on the diffusive limitation imposed on evapotranspiration or in the relative
utilization of RuBPC and PEPC in CAM, and generally in the use of water.
All this is reflected in isotope composition and thus, various **ecotypes of
plants**, like halophytes and xerophytes can be differentiated.

Another example from metabolism is represented by **N-nutrition** of
plants. N-salts in the soil tend to enrich the heavier isotope ^{15}N, as compared to atmospheric gaseous N_2, and therefore one can recognize symbiotic N_2-fixers (e.g. legumes with root nodules) by the lower ^{15}N content.
Analyses of the natural abundance of ^{15}N in soils have also served to document forest-to-pasture chronologies and record changes of land-use pattern
in the western Amazon Basin in Brazil (Piccolo et al. 1994).

The flow of various isotopes through the biomass of plants also affects the
transfer into other **compartments of ecosystems.** This allows the study of food
webs, habitat preferences in wandering animals, and even the analysis of eating and drinking habits in human populations and individuals. For the latter it
usually suffices to analyze the organic matter of hairs or finger- and toenails.

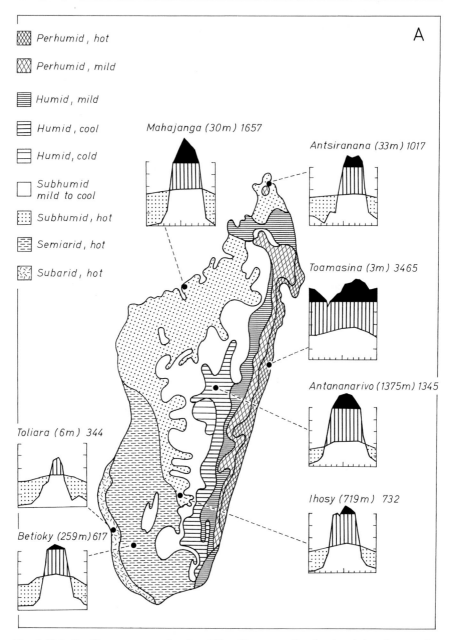

Fig. 2.13A, B. Climate zones related to *Klimadiagramm* distribution (**A**) and vegetation types (**B**) of Madagascar. The vegetation map (**B**) contains points indicating ranges of $\delta^{13}C$ values as explained in the *inset*. Note that the *closed symbols* marking the more negative $\delta^{13}C$ values are concentrated in the wetter regions, and the *open symbols* marking less negative $\delta^{13}C$ values are accumulated in the drier vegetation units. The *inset* also gives the frequency of $\delta^{13}C$ values for the samples collected on the island for three combinations of vegetation units as indicated. (After Kluge et al. 1991)

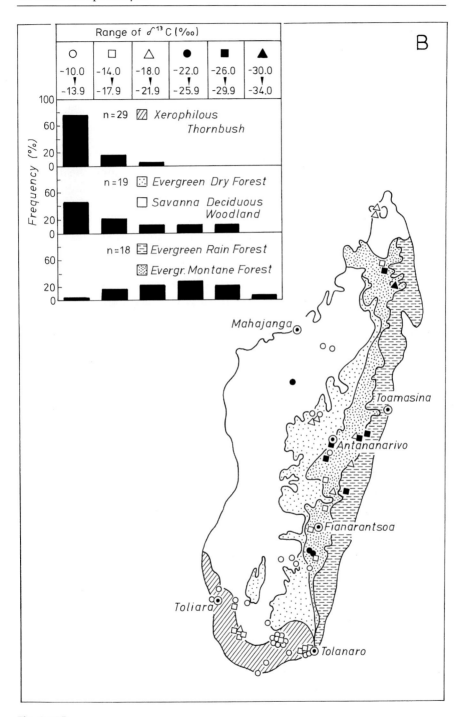

Fig. 2.13B

Since the analysis of dry matter of the organisms is often sufficient, one can cover large geographic areas with sampling even from remote regions using simple equipment. Even collections in herbaria may be used. In this way, for example one can arrive at conclusions about the **large-scale ecological distribution of modes of photosynthesis.** C_4-grasses dominate in tropical savannas, their relative abundance declines with increasing altitude (Tieszen et al. 1979; Medina 1982, see Sect. **7.2.1**).

An impressive example, if not for a whole continent, is given for the rather large tropical island of Madagascar (Kluge et al. 1991). Combining the *Klimadiagramm* method with the stable-isotope technique, the distribution of CAM among species of the genus *Kalanchoë* has been studied and related to climatic zones and vegetation types of the island. There are 52 species of *Kalanchoë* in Madagascar, of which all are either obligate or facultative CAM plants. There is high flexibility among the species to obtain a variable amount of carbon by direct CO_2 fixation via RuBPC, and this is reflected by increasingly negative $\delta^{13}C$ values, whereas primary CO_2-fixation dominated by PEPC leads to less negative $\delta^{13}C$ values. The large scale effects deduced from the analysis present a very close correlation of $\delta^{13}C$ values in the dry matter of *Kalanchoë* species and climate and vegetation zones on the island. Less negative (CAM-like) values are dominant in the drier zones with evergreen dry forest, deciduous woodland, savannas and xerophilous thornbush, while more negative (C_3-like) values prevail in the wetter zones with evergreen rainforest and montane forest (Fig. **2.13**). The example illustrates the close relations between climate, vegetation types and prevalence of the water conserving CAM-mode of photosynthesis in a given genus.

Large-scale isotope effects, however, may also result from transfer rates. Thus, respiration and photosynthesis of organisms determine the vertical $^{13}CO_2$-gradient in tropical rainforests (Medina et al. 1986). Richey et al. (1990) have proposed to use ^{18}O-analyses to assess the large-scale consequences of the destruction of tropical rainforest in the Amazon region. Since C-, N- and S-compounds from biogenic and anthrogenic sources have different isotope composition, the pathways of pollutant emission can also be traced. The large global-cycles of a range of elements in the gaseous atmosphere, as well as regional environmental effects, can be diagnosed in this way.

Finally, interest has recently been focussed on the element **strontium**, which is not subject to metabolism, in terms of **mineral nutrition.** The $^{87}Sr/^{86}Sr$ ratios in the bed-rock develop in geological times. The different geochemical mobilization of the isotopes allows distinctions between nutrient supply of vegetation from the soil, water or dust in the atmosphere. This facilitates understanding of nutrient sources in ecosystems (Graustein 1989).

2.6
Data Analysis

The huge amounts of data which become available in large-scale diagnostics require hardware and software technologies for storage, organization and analysis of data. Special mathematical and statistical procedures are essential to reduce the relevant information from the mass of data to an interpretable form. We deal with multidimensional problems, which have to be summarized numerically or graphically and must be cast into a comprehensive presentation. Special geographic information systems with multifactorial mapping are designed to provide surveys related to space (Wallace and Campbell 1990).

This may possibly constitute the most important bottleneck. Hobbs and Mooney (1990) conclude their book on remote sensing of the biosphere functions with the statement, that even without the intensive further development of new sensors, the currently available technologies offer so much that the capacity for interpretation and application has been surpassed. The major problems do not appear to be with collection but with analysis and understanding of data.

References

Chapelle EW, Wood FM, McMurtrey JE, Newcomb WW (1984a) Laser-induced fluorescence in green plants: 1. A technique for the remote detection of plant stress and species differentiation. Appl Opt 23:134–138

Chapelle EW, McMurtrey JE, Wood FM, Newcomb WW (1984b) Laser-induced fluorescence of green plants. 2. LIF caused by nutrient deficiencies in corn. App Opt 23:139–142

Ehleringer JR, Rundel PW (1989) Stable isotopes: history, units, and intrumentation. In: Rundel PW, Ehleringer JR, Nagy KA (eds) Stable isotopes in ecological research. Ecological studies, vol 68. Springer, Berlin Heidelberg New York, pp 1–19

Ehrendorfer F (1991) Geobotanik. In: Sitte P, Ziegler H, Ehrendorfer F, Bresinsky A (eds) Strasburger Lehrbuch der Botanik. G Fischer, Stuttgart

Gates M (1965) Energy, plants and ecology. Ecology 46:1–14

Gosz JR, Dahm CN, Risser PG (1988) Long-path FTIR measurement of atmospheric trace gas concentrations. Ecology 69, 1326–1330

Graustein WC (1989) $^{87}Sr/^{86}Sr$ ratios measure the sources and flow of strontium in terrestrial ecosystems. In: Rundel PW, Ehleringer JR, Nagy KA (eds) Stable isotopes in ecological research. Ecological studies, vol 68. Springer, Berlin Heidelberg New York, pp 491–512

Hák R, Lichtenthaler HK, Rinderle U (1990) Decrease of the chlorophyll fluorescence ratio F690/F730 during greening and development of leaves. Radiat Environ Biophys 29:329–336

Hobbs RJ, Mooney HA (eds) (1990) Remote sensing of biosphere functioning. Ecological studies, vol 79. Springer, Berlin Heidelberg New York

Hoge FE, Swift RN, Yungel JK (1983) Feasibility of airborne detection of laser-induced fluorescence emissions from green terrestrial plants. Appl Opt 22:2991–3000

Jones PD (1990) Le climat des mille dernières années. Recherche 21:304–312

Kluge M, Brulfert J, Ravelomana D, Lipp J, Ziegler H (1991) Crassulacean acid metabolism in *Kalanchoë* species collected in various climatic zones of Madagascar: a survey by $\delta^{13}C$ analysis. Oecologia 88:407–414

Lichtenthaler HK, Hák R, Rinderle U (1990) The chlorophyll fluorescence ratio F690/F730 in leaves of different chlorophyll content. Photosynth Res 25:295–298

Malingreau J-P, Tucker CJ (1987) La végétation vue de l'espace. Recherche 18:180–189

Matson PA, Harriss RC (1988) Prospects for aircraft-based gas exchange measurements in ecosystem studies. Ecology 69. Ecol Soc of Am, Washington, DC (5), 1318–1325

Matson PA, Vitousek PM (1990) Remote sensing and trace gas fluxes. In: Mooney HA, Hobbs RJ (eds) Remote sensing of biosphere functioning. Springer, Berlin Heidelberg New York, pp 157–167

Medina E (1982) Physiological ecology of neotropical savanna plants. In: Huntley BJ, Walker BH (eds) Ecology of Tropical Savannas, Ecological Studies, vol 42. Springer, Berlin Heidelberg New York, pp 308–335

Medina E (1986) Forests, savannas and montane tropical environments. In: Baker, NR, Long SP (eds) Photosynthesis in contrasting environments. Elsevier, Amsterdam, pp 139–171

Medina E, Montest G, Cuevas E, Roksandic Z (1986) Profiles of CO_2 concentration and $\delta^{13}C$ values in tropical rain forests of the upper Rio Negro basin, Venezuela. J Trop Ecol 2:207–217

Nobel PS (1983) Biophysical plant physiology and ecology. WH Freeman, San Francisco

Piccolo MC, Neill C, Cerri CC (1994) Natural abundance of ^{15}N in soils along forest-to-pasture chronosequences in the western Brazilian Amazon Basin. Oecologia 99:112–117

Richey JE, Adams JB, Victoria RL (1990) Synoptic-scale hydrological and biogeochemical cycles in the Amazon river basin: a modelling and remote sensing perspective. In: Hobbs RJ, Mooney HA (eds) Remote sensing of biosphere functioning. Ecological studies, vol 79. Springer, Berlin Heidelberg New York, pp 249–268

Rundel PW, Ehleringer JR, Nagy KA (eds) (1988) Stable isotopes in ecological research. Ecological studies, vol 68. Springer, Berlin Heidelberg New York

Running SW (1990) Estimating terrestrial primary productivity by combining remote sensing and ecosystem simulation. In: Hobbs RJ, Mooney HA (eds) Remote sensing of biosphere functioning. Ecological studies, vol 79. Springer, Berlin Heidelberg New York, pp 65–86

Sellers PJ, Hall FG, Strebel DE, Asrar G, Murphy RE (1990) Satellite remote sensing and field experiments. In: Hobbs RJ, Mooney HA (eds) Remote sensing of biosphere functioning. Ecological studies, vol. 79. Springer, Berlin Heidelberg New York, pp 169–201

Smith BN (1975) Carbon and hydrogen isotopes of sucrose from various sources. Naturwissenschaften 62:390

Tieszen LL, Senyimba MM, Imbamba SK, Troughton JH (1979) The distribution of C_3 and C_4 grasses and carbon isotope discrimination along an altitudinal and moisture gradient in Kenya. Oecologia 37:337–350

Wallace JF, Campbell N (1990) Analysis of remotely sensed data. In: Hobbs RJ, Mooney HA (eds) Remote sensing of biosphere functioning. Ecological studies, vol 79. Springer, Berlin Heidelberg New York, pp 291–304

Walter H (1973) Vegetationszonen und Klima. Ulmer, Stuttgart

Walter H (1982) Bekenntnisse eines Ökologen. 3. Auflage. G Fischer, Stuttgart

Walter H, Breckle S-W (1983) Ökologie der Erde, Bd 1, Ökologische Grundlagen in globaler Sicht. G Fischer, Stuttgart

Walter H, Breckle S-W (1984) Ökologie der Erde, Bd 2, Spezielle Ökologie der tropischen und subtropischen Zonen. G Fischer, Stuttgart

Walter H, Lieth H (1967) Klimadiagramm – Weltatlas. G Fischer, Jena

Wessman CA (1990) Evaluation of canopy biochemistry.In: Hobbs RJ, Mooney HA (eds) Remote sensing of biosphere functioning. Ecological studies, vol 79. Springer, Berlin Heidelberg New York, pp 135–156

Woodward FI (1987) Climate and plant distribution. Cambridge University Press, Cambridge

Ziegler H (1989) Hydrogen isotope fractionation in plant tissues. In: Rundel PW, Ehleringer JR, Nagy KA (eds) Stable isotopes in ecological research. Ecological studies, vol 68. Springer, Berlin Heidelberg New York, pp 105–123

CHAPTER 3

Tropical Forests

Usually the term "tropical forest" raises the association with "rainforest", and indeed, as noted above (Sect. **1.4**), the current world-wide debate on the threat to the tropical environment is largely focused on the "tropical rainforest". However, the question arises: What do we define as "tropical rainforest"? The answer is not simple, and it turns out that we must work with a distinction of different types of tropical forest as developed in the first two Sects. of this Chap. (Sects. **3.1** and **3.2**). However, notwithstanding the noticeable differences in the physiognomy of the various types of tropical forests, they have much in common independent of whether they are evergreen, seasonal or semi-evergreen. Thus, some theoretical considerations regarding the structure of forests can be made for tropical forests in general (Sect. **3.3**). This also applies to the discussion of environmental factors, such as minerals in the soil (Sect. **3.5**) and light (Sect. **3.6**). However, at the lowest level of water availability still allowing the formation of forest, a type of tropical forest develops which is so different in both physiognomy and the requirements of stress responses leading to very specific physiological and biochemical adaptations, that it is covered in a separate Sect., namely the thornbush-succulent forest (Sect. **3.7**). Finally, mangroves constitute a type of tropical forest which is uniquely determined by salinity and soil inundation along marine coastlines, so that they are treated in a separate Chap. (Chap. **5**).

3.1
Tropical Rainforest: Definition and Distribution

The problem of deciding what we define as "tropical rainforest" is best illustrated by eight maps of Venezuela presented by Vareschi (1980), where he depicts the distribution of rainforest in this tropical country according to the views of different authors (Fig. **3.1**). In the two extreme cases either almost 2/3 of the whole country is covered by rainforest (upper left in Fig. **3.1**) or there is no rainforest at all (lower right in Fig. **3.1**); and there are gradations in between these extremes.

The purist's definition of tropical rainforest requires that there should be no **seasonality** whatsoever. In the tropics seasonality means unequal distri-

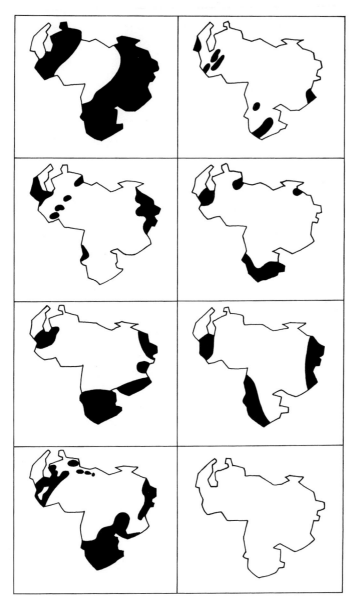

Fig. 3.1. Distribution of "rainforest" (*black*) in the tropical country Venezuela according to different authors. (Vareschi 1980, with kind permission of R. Ulmer)

bution of rainfall over the year. Fig. **3.2** shows that seasonality in rainfall is absent over only a very narrow zone 1° north and south of the equator. Hence, the lack of any rainforest in Venezuela accords with the purist's definition (Fig. **3.1** lower right). If one begins to broaden the definition, of

Fig. 3.2.
Dry and wet seasons in the subtropical and tropical zone with zenithal precipitation. Wet seasons *hatched*, dry seasons *dotted*. (Walter and Breckle 1984, with kind permission of S. -W. Breckle and G. Fischer-Verlag)

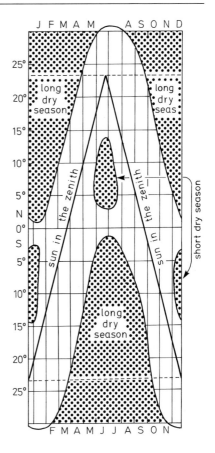

course, it becomes a matter of taste how far this term may be extended. How long should rainless periods be and in how many successive years should they occur in order to retain the term "rainforest"? Accordingly, we then derive the gradations seen in the maps of Venezuela, shown in Fig. **3.1.**

In addition to seasonality there are other features which characterize tropical rainforests. One of the most conspicuous is the extraordinary **diversity of tree species.** In contrast to the temperate and boreal zones, where forests can be named after dominant tree species, e.g. spruce, fir, pine, beech, oak, birch etc., there is no dominance of anyone particular tree species in tropical rainforests. One may encounter up to 300 tree species per ha, which represent about 1/3 of all plant species present. The most frequent tree species rarely represent more than 15 % of all species of trees present (Jacobs 1988; Whitmore 1990). This phenomenon also makes it difficult to define the **minimal quadrat** of sampling plots in plant sociology. The minimal quadrat is given by the size of a plot in a system being studied above which the total number of species observed does not increase (see Sect.

3.3.1 below, Fig. **3.11**). Such minimal quadrats in tropical rainforests may be quite large and may never be attained.

Thus, tropical forests provide a good illustration for the problem given by abstraction in nature. "To realize abstraction in reality means to destroy reality" (Hoffmeister 1955, after Vareschi 1980). The only way out of the dilemma is to distinguish different types of tropical forest.

3.2
Types and Physiognomy of Tropical Forests

It is clear from the above that **rainfall** must be the most important factor determining the definition of different types of tropical forest. Gradation of rainfall is inherent in Beard's (1946, 1955) distinction of forests according to **altitude** (Fig. **3.3**):

- low-land rainforest,
- lower montane rainforest,
- upper montane rainforest,
- elfin forest.

A much more detailed separation is obtained when forest types are related to *Klimadiagramm* graphs (after Walter 1973, see Sect. **2.2.1**, Box **2.2**), which depict the duration and severeness of dry and wet seasons. Using India and Venezuela as examples comparison of forest types is made with a diagram of precipitation versus extension of dry periods (Fig. **3.4**). The corresponding *Klimadiagramm* graphs for Venezuela are shown in Fig. **3.5**.

This type of diagram allows a clear separation of various types of tropical forest according to the degree of seasonality which increases with increasing numbers of dry months per year and decreasing annual rainfall:

- evergreen rainforest,
- seasonal rainforest,
- semi-evergreen moist and dry monsoon or trade-wind forest,
- drought-deciduous forest, and
- thorn scrub.

The term cactus-forest was coined by Vareschi (1980) and is further discussed below (Sec. **3.7**). An example for the physiognomy of a semi-seasonal rainforest is given in Fig. **3.3A**, the drier types, i.e. drought deciduous forest, thorn scrub and cactus-forest are shown in Fig. **3.6** and below in Figs. **3.51**, **3.52** and **3.53**.

Humid and arid climates, respectively, are defined by the difference between **rainfall and evaporation.** Positive values indicate a humid climate, where precipitation is larger than evaporation, and negative values mark arid climates with evaporation larger than precipitation. As we have

Fig. 3.3A-D. Types of tropical forest at different altitudes. **A** Semi-evergreen lowland rainforest (East Venezuela). **B** Montane rainforest (northern range Trinidad). **C** Upper montane rainforest (cloud or fog forest; Rancho Grande, northern coastal range Venezuela). **D** Elfin forest (Serro Santa Ana, Paraguana Peninsula Venezuela)

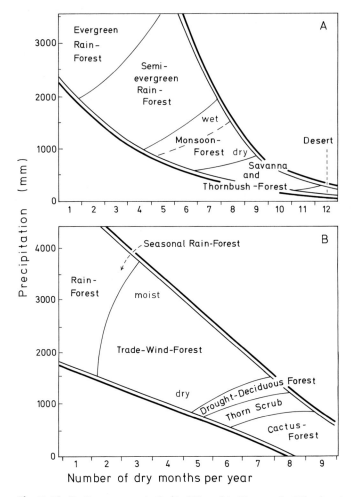

Fig. 3.4A, B. Forest types in India (**A**) and in Venezuela (**B**) related to annual amount of precipitation and the duration of drought periods (number of dry months per year). (After Walter and Breckle 1984 and Vareschi 1980, with kind permission of R. Ulmer)

demonstrated above (Sect. **2.2.1**), the *Klimadiagramm* technique based on temperature and precipitation is highly successful in demonstrating the degree of aridity and humidity of individual stations as well as large areas. In contrast to data on temperature and precipitation, which usually are readily available from many weather stations, measurements of free evaporation are scarce. Other factors, especially wind, may significantly affect evaporation. Thus, where evaporation data is available, it can add much to the information provided by *Klimadiagramm* graphs. As seen in the examples of 7 stations in Venezuela (Fig. **3.7**) the results of *Klimadiagramm* graphs regarding the occurrence of humid seasons is roughly confirmed by

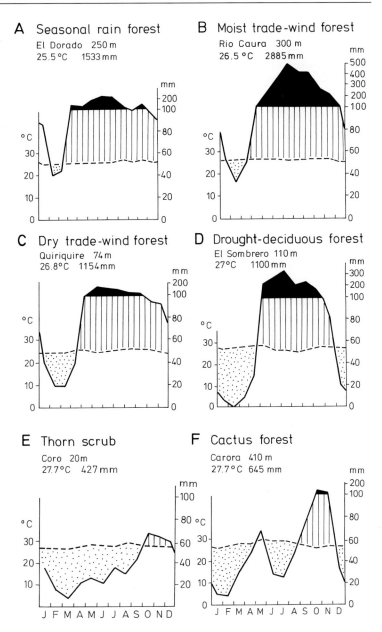

Fig. 3.5A-F. *Klimadiagramm* graphs for the different forest types distinguished in the diagram of Fig. **3.4B**

Fig. 3.6A, B. Types of dry tropical forest. **A** Drought deciduous forest (Falcon, Vene-zuela) **B** Cactus-forest (Carora, Venezuela)

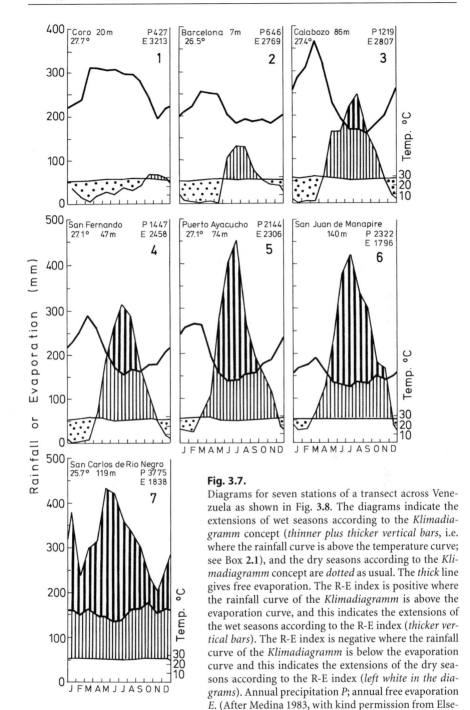

Fig. 3.7.
Diagrams for seven stations of a transect across Venezuela as shown in Fig. **3.8**. The diagrams indicate the
extensions of wet seasons according to the *Klimadiagramm* concept (*thinner plus thicker vertical bars*, i.e.
where the rainfall curve is above the temperature curve;
see Box **2.1**), and the dry seasons according to the *Klimadiagramm* concept are *dotted* as usual. The *thick* line
gives free evaporation. The R-E index is positive where
the rainfall curve of the *Klimadiagramm* is above the
evaporation curve, and this indicates the extensions of
the wet seasons according to the R-E index (*thicker vertical bars*). The R-E index is negative where the rainfall
curve of the *Klimadiagramm* is below the evaporation
curve and this indicates the extensions of the dry seasons according to the R-E index (*left white in the diagrams*). Annual precipitation *P*; annual free evaporation
E. (After Medina 1983, with kind permission from Elsevier Science-NL, Sara Burgerhartstraat 25, NL-1055 KV
Amsterdam, The Netherlands)

Fig. 3.8.
Transect across Venezuela with rainfall minus evaporation (R-E index) at different stations with various forest types. *Numbers in parentheses* indicate dry months per year after the R-E index. See also *Klimadiagramm*s in Fig. **3.7**. (After Medina 1983, with kind permission from Elsevier Science-NL, Sara Burgerhart-straat 25, NL-1055 KV Amster-dam, The Netherlands)

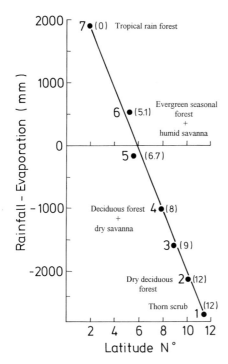

the **rainfall minus evaporation (R-E) index.** However, the wet seasons generally appear somewhat shorter with the R-E index, and on the basis of this criterion small apparent humid seasons of the *Klimadiagramm* graphs disappear entirely (Fig. 3.7 diagrams 1 and 2). These 7 stations represent a latitudinal transect from the wet continental areas close to the equator to the drier coastal regions in the north of Venezuela (Fig. **3.8**). In addition to the

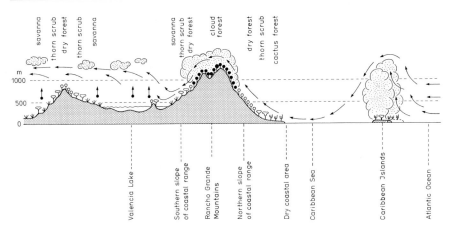

Fig. 3.9. Profile in a NW to SE direction (*from right to left*) explaining the formation of cloud forest on the northern coastal range of Venezuela. (After Vareschi 1980, with kind permission of R. Ulmer)

effect of **altitude** stressed by the model of Beard a very strong effect of **latitude** on the expression of tropical forest types is highly evident, which agrees with the predictions of Fig. **3.2**.

Among the forests depicted in Figs. **3.3** and **3.6** the two extreme cases, i.e. the upper montane fog or cloud forests (Fig. **3.3C**) and the cactus-forest (Fig. **3.6B**), deserve some special comments here. Other types of forests like mangroves and gallery forests are dealt with in a separate chapter (Chap. **5**) and mentioned again in relation to savannas (Chap. **7.1**) respectively.

One of the best preserved tropical **cloud forests** is the forest of Rancho Grande on the northern coastal range of Venezuela. It is determined by the trade winds coming from a north-eastern direction (Fig. **3.9**). Clouds build up over the Caribbean islands and dissolve as the wind moves south, where it then hits the coastal range of mountains on the South American continent. The hot wind cools down as it climbs upwards and on top clouds are formed almost continuously (Figs. **3.3C** and **3.9**). We find an altitudinal gradation of vegetation from cactus-forest, thorn scrub and dry forest to evergreen cloud forest. Clouds break up as the wind drives down the southern slopes and also move further south with the trade winds. In places, the constant exposure to the wind may lead to formation of **elfin forest** with dwarf trees similar to those on the top of Santa Ana Mountain on the Paraguana Peninsula of Venezuela, which are here dominated by *Clusia multiflora* and the palm *Geonema paraguanensis* about 1 m tall (Fig. **3.3D**).

The term **cactus-forest** sensu Vareschi (1980) (Fig. **3.6B**) relies on the acceptance of large columnar cacti as tree-like plants, and depends on the size of the area occupied by them. Alternatively or additionally, one may require the most important physiognomical aspect of a forest to be the formation of a closed canopy. The area around Carora in Venezuela, for which Vareschi originally coined the term, is dominated by columnar cacti of *Cereus lemairei*, *Ritterocereus griseus* and *Cephalocereus moritzianus*. However, it appears, that woody Mimosaceae, Capparidaceae and Caesalpiniaceae like *Cercidium praecox* very much add to the impression of a closed canopy, which is certainly given when one is walking around in these forests. The floor of these forests is bare of vegetation or covered by a forbidding muddle of thorny cacti, particularly *Opuntia wentiana* and *O. caribea*. This may be due to a large extent to overgrazing by goats. At places where the access of goats is prevented, grasses are entering these cactus-forests, which then obtain a physiognomy more comparable to the equivalent thornbush-savanna in Africa (see also Sect. **3.7**).

3.3
Structure of Tropical Forests

There are three dimensions to consider in the structure of tropical forests: while we may distinguish between the extent of communities, ecosystems and landscapes in space largely by applying a two dimensional approach, horizontal stratification adds the third dimension. The first problem, i.e. the horizontal structure of tropical forests (Sect. **3.3.1**) leads us into consideration of more theoretical questions such as **succession, plasticity** and **diversity**. Conversely, stratification, meaning the vertical structure of tropical forests (Sect. **3.3.2**), is much more directly linked to local action of **specific environmental factors,** such as light, CO_2 and minerals, whose vertical distribution can be described.

3.3.1
Horizontal Structure

3.3.1.1
Diversity: The Spatial Structure of the Environment

The very high diversity of tropical forests has been alluded to above with respect to floristic diversity within a given type of forest (Sect. **3.1**) and diversity of various types of forest (Sect. **3.2**).

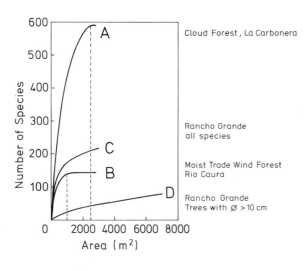

Fig. 3.10. Species/area curves for four tropical forests in Venezuela. *A* Cloud forest La Carbonera (minimal quadrat 2500 m²); *B* moist trade-wind forest Rio Caura (minimal quadrat 1100 m²); *C* and *D* cloud forest of Rancho Grande, C all species, D only trees with a stem diameter > 10 cm, for which the minimal quadrat has not been attained at 7500 m². (After data of Vareschi 1980, from Lüttge 1995)

Fig. 3.11.
Floristic diversity related to species-area diagrams. (After Whithmore 1990, from Lüttge 1995). $\alpha\beta$.... Species-poor community (low α-diversity) with low β-diversity, *viz.* few species, small minimal quadrat; $\alpha\beta$.... species-poor community (low α-diversity) with high β-diversity, *viz.* few species, large minimal quadrat; $\alpha\beta$.... species-rich community (high α-diversity) with low β-diversity, *viz.* many species, small minimal quadrat; $\alpha\beta$.... species-rich community (high α-diversity) with high β-diversity, *viz.* many species, large minimal quadrat; γ.... bi- (or multiphasic) curves indicating γ-diversity

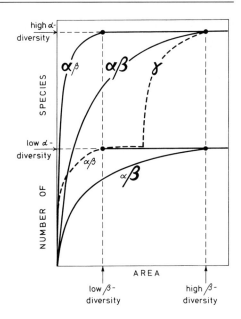

The high floristic diversity in tropical forests leads to the problem that the **minimum quadrats** of plant sociology become very large. Minimum quadrats give the smallest possible area in which all species occurring in a habitat are present, as illustrated in Fig. **3.10**. Whitmore (1990) has used such species to area diagrams to illustrate the concept of **floristic α-, β- and γ-diversity** referring to different levels in landscapes (Noss 1983) (Fig. **3.11**) as first defined by Whittaker (1975), i.e. by taking the **richness of species** as given by the number of different species per minimum quadrat

- **at the level of communities** or coenoses, α-diversity,
- **at the level of ecosystems** with different vegetation types, for example at ridges, on hillsides, in valleys etc., β-diversity, and
- **at the level of landscapes** comprising several ecosystems, γ-diversity.

An other approach is the evaluation of diversity in the spatial structure of the physical environment in relation to **environmental factors**. Bell et al. (1993) have suggested to use log-log regressions of environmental factors, such as edaphic variables, water chemistry and climate, *versus* distance over large spatial scales (10^6 m). Plotting the log variance of environmental factors, e.g. the soil nitrogen, soil phosphorus etc., on log distance allows comparisons of both the heterogeneity of different environments with respect to that given factor and the different factors within a particular environment. With consideration of environmental factors, of course, **ecophysiological aspects are introduced into the discussion of diversity.** Varying ecophysiological behaviour of given genotypes is expressed as phenotypic plasticity. In the log-log analysis of regression of variance of environmental factors on distance by Bell et al. (1993) there was no indication that the variance of the physical environment tended to approach some maximal value as the distance increased. In contrast, it increased continuously with distance. Thus, the slopes of the log-log regressions provide a means to compare heterogeneity of environments and factors in relation to each other, i.e. slopes of the log-log regressions indicate the correlations of nearby sites between each other. Where nearby sites are highly correlated, selection will tend to favour specialization because dispersing offspring find conditions for growth similar to their parents. This leads to diversity of genotypes. On the other hand, where nearby sites show little correlation between each other, offspring tend to find conditions different to their parents. This favours plasticity.

3.3.1.2
Plasticity, Diversity and the Biological Stress Concept

Environmental factors direct us to the consideration of the role of stress, because any environmental factor can become a **stress factor** or **stressor** if its dosage is too high or too low. This is explained by the biological stress concept as described in Box **3.1**. We may ask the questions as to how stress may be involved in regulating plasticity and diversity and whether in fact, plasticity and diversity are related.

In an experiment applying different degrees of stress to experimental microcosms, Grime et al. (1987) have demonstrated that high diversity is given only within a rather narrow range of stressed conditions. For conditions in the British Isles high species diversity occurred at stress intensities which allowed no less than 350 g dry biomass m^{-2}, but no more than 750 g m^{-2}. At lower stress there is dominance of one or only a few robust and competitive species. At higher stress only a few highly adapted "specialists" survive. In a Malaysian rainforest it was shown that the species diversity of trees and vines strongly responded to combined phosphorus and potassium concentrations in the soil. Diversity was highest on soils with a low P + K

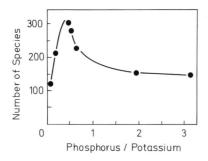

Fig. 3.12.
Relationship between the diversity of trees and vines in a Malaysian rainforest (*ordinate*) and combined phosphorus and potassium concentrations in the soil (*P + N index on the abscissa*). (After Tilman 1982)

Box 3.1

The biological stress concept

A good physical analogue for developing the terminology is a spring.
Stress is put on the spring by **strain**. **Reversible stress** is brought
about by strain in the elastic range of the spring material: **elastic strain**.
Irreversible stress is due to strain beyond the elastic range of the
spring material: **plastic strain**.

The biological stress concept was developed by Selye (1973), Levitt
(1980) and Larcher (1987). Any external factor (biotic or abiotic) and
internal factor can induce stress, i.e. become a **stressor**, if its dosage is
too high or too low. The terminology of the biological stress concept is
explained in the diagram giving four different possible cases for the
development of a biological system with time (abscissa).

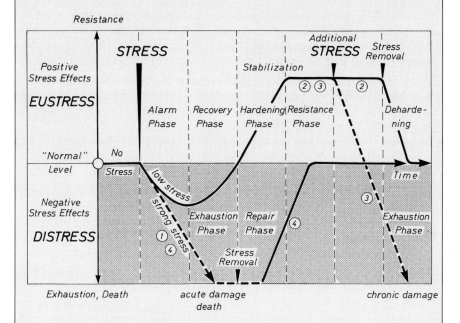

1. *Strong stress*
   Strong stress out of an **alarm phase** more or less rapidly leads into a
   phase of **exhaustion** followed by **acute damage** and death. The
   stress has negative effects, it is a **distress**.

Box 3.1 (Continued)

2. *Low stress followed by stress removal*
   Low stress leads into an **alarm phase** generating recovery mechanisms. In a **recovery phase** the system develops out of the conditions in which stress has negative effects, to conditions in which stress
   has positive effects and stimulates the system; stress is a **eustress**.
   The system stabilizes during a **hardening phase** and attains a **resistance phase**, in which it may remain, unless the degree of stress is
   changed or **external or internal reserves** required for resistance are
   exhausted. With respect to the latter, it is clear that time, i.e. the duration of stress application may be important. Upon stress removal the
   system enters a **dehardening phase** and returns to the normal level.

3. *Low stress followed by additional stress*
   The system first develops like that of case (2), but then additional
   stress is applied either by the original stress becoming stronger or by
   additional different stressor(s). The system now goes into the condi-
   tion of distress, an exhaustion phase and **chronic damage**.

4. *Strong stress with acute damage followed by repair after stress removal*
   The system first develops like that of case (1), but then a stress-free
   period follows, and during a **repair phase** the system is restored and
   returns to the "normal" level.

(After Beck and Lüttge 1990.)

index, it declined towards both very poor soils and soils with more ample
nutrient supply (Fig. **3.12**). The relationship between **diversity** and environmental richness can also be demonstrated for tropical birds and mammals
(Reichholf 1994).

With regard to **plasticity** it appears that both high stress and low stress
do not favour traits, which support such phenotypic variability. **High stress
favours specialized adaptations** to the prevailing specific and strong stress-
or (e.g. frost near the poles or drought in deserts). **Low stress allows the
success of few species**, which can competitively procure resources for development of their own biomass (e.g. nitrophilous plants in sites rich in nitrate
and other nutrients). Only **medium stress advances the unfolding of variability**. Stress of medium intensity and high variability in time is typical for
the environment of the tropical forests, where the important factors are

– nutrients,
– water,

– CO_2,
– light,
– temperature.

None of these factors ever really becomes extreme, but their variations and interactions will cause stress in tropical forests as described in the later Sects. of this Chap..

Plasticity itself can be considered as a trait, which is subject to selection (West-Eberhard 1986, 1989). Since selection is acting on the phenotypes, **phenotypic plasticity offers material for selection**. This is particularly important in systems, which are not strictly homeostatic (see below: Sect. **3.3.1.3**), and where radiative movement of phenotypes followed by separation may be **one of the bases for the development of genetic diversity** (West-Eberhard 1986, 1989). Thus, the promotion of phenotypic plasticity by variable and medium stress may be one of the reasons for the extraordinarily high biodiversity of tropical forests (Lüttge 1995).

3.3.1.3
Successions: Homeostatic Climax Associations or Deterministic Chaos of Oscillating Mosaics

After assessing diversity and plasticity, this then leads us to the third question: Is there **homeostasis**? The tropical rainforest is frequently considered to be a **climax association**. Ideally, climax associations are steady-states for vegetation, representing stable or homeostatic ecological equilibria, determined by the natural environmental factors in a given climatic zone. According to the climax theory (Clements 1936) independent of the starting conditions at a given location, **progressive successions** should always lead to the same final equilibrium or climax association. Only when there is an effect of external influences, such as natural or man-made catastrophes, are **regressive successions** elicited, which cause a deviation from the climax association. Of course, the latter involves changes of environmental conditions.

The question arises, however, as to whether the final steady state is really independent of the starting conditions. Can predictions be made? Is there only one possible final equilibrium? Or are there several different possible ecological states? Is it then possible that small initial deviations from the mean climatic conditions (e.g. mean annual precipitation and temperatures) might divert the development of succession towards very different end points?

In fact, the straightforward application of the climax theory to large ecosystems has been challenged by Remmert (1985, 1991), using the very example of the tropical rainforest. He strongly underlines the dynamic nature of rainforests (see also Whitmore 1990) and in his view the tropical rainforest is subject to a continuous cycle of series of successional states. It

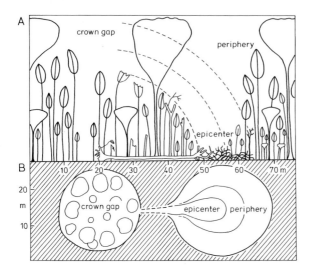

Fig. 3.13A, B. Formation of gaps or chablis by falling trees. Subsequently the crown gap fills more rapidly than the gap created by the impact. (After Jacobs 1988). **C** Fallen tree in a forest of Sierrania Páru, Venezuela

represents a diverse cyclically changing mosaic pattern (see also Watt 1947). This can be illustrated by considering the dynamics of **chablis** in the tropical rainforest. In the original French meaning chablis are clearings in forests due to storms. In wet tropical forests tall falling trees with their large crowns cause two adjacent gaps, one beneath the original location of the crown and the other one at the site of impact on the ground (Fig. **3.13**).

When this is set in the context of floristic diversity, destruction such as the formation of chablis, as well as the introduction of roads may increase diversity because β-diversity is introduced (Sect. **3.3.1.1**). Gaps and chablis are reinvaded by vegetation, and with various successional stages the forest is restored (Fig. **3.14**). Thus, such chablis are sites of destruction and renewal, which at any given time may comprise from 3 to 10 % of the total forest area (JACOBS 1988). Larger clearings also result from shifting agriculture (Fig. **3.15**) and other human activities. There is no pre-determined pattern of renewal of forest in natural clearings resulting from falling trees, hurricanes, earthquakes, volcanic eruptions, fires and landslides, or in farms abandoned due to exhaustion of nutrients or the take-over of weeds and pests. As Jacobs (1988) comments "... the selection is one of unpredictable irregularity", and depending on the adjacent vegetation, the extent of diversity in the forests and the unpredictability of proximal species, regeneration will be related to the availability and viability of propagules and the age of the surrounding communities.

Such unpredictability is generally interpreted in phytosociology as demonstrating that the distribution of plants and the development of diversity is stochastic. Taking the term stochastic in its strict mathematical meaning, this is by no means justified. Stochastic white noise in empirical time series can not readily be distinguished from the so-called deterministic chaos, which follows strict mathematical rules. A distinction between the two can only be made via sophisticated theoretical analyses requiring very detailed sets of time series data. These are rarely available, and hence, it is hard to prove whether deterministic chaos or stochastic noise predominate.

Nevertheless, it is important to consider briefly the possible implications of the theory of deterministic chaos in ecology. Due to the pioneering contributions of R. May (1976) population dynamics has become one of the roots of the development of the chaos theory. The apparently simple logistic equation discussed by May (1976) also gave biologists ready access to an intuitive comprehension of the theory of deterministic chaos. Let x_t be the size of a population at a certain state at time t. We then want to know the size of the population at the next possible state in time, i.e. x_{t+1}. It is obvious that the development of the population depends on its resources. These may be given by a growth factor $\underline{r}$ or more generally an external control parameter, such that x_{t+1} is proportional to $\underline{r} \cdot x_t$. However, it is not only evident from the exorbitant increase of the human population on the globe, but a general experience of population ecology, that increasing population den-

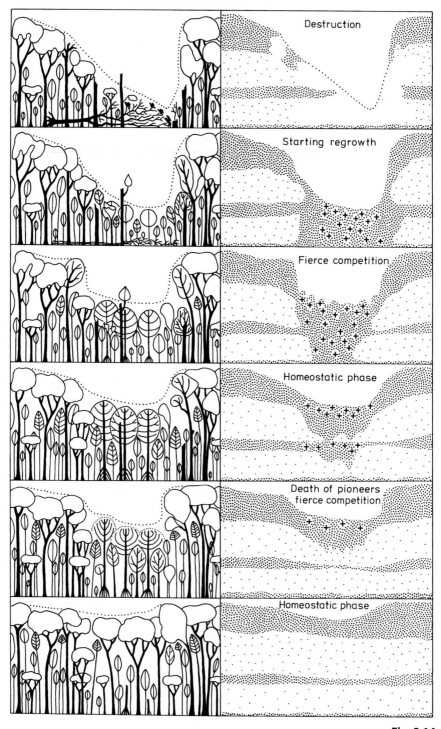

Fig. 3.14

Fig. 3.15. Slash and burn agriculture (see also Fig. 2.5)

◀ **Fig. 3.14.** Reinvasion of a chablis. (After Jacobs 1988)

sities also bear inhibitory mechanisms in themselves, i.e. x_{t+1} is proportional to $x_t \cdot (1-x_t)$. Hence, the logistic equation for the development of the population x is

$$x_{t+1} = r \cdot x_t \cdot (1 - x_t).\qquad(3.1)$$

Subsequent population sizes can be calculated by recursions, where x_{t+1} is used in the place of x_t to obtain x_{t+2} and so forth.

However, the equation is only apparently simple. It describes one of several possible routes from order or regularity into chaos (Schuster 1995; Fig. **3.16**). It shows that ordered predictability only occurs for a narrow range of the value of the control parameter r. At low values of r there is a steady state, while at larger values of r a bifurcation (i.e. a branching or dichotomy) leads first to phase doubling and ordered oscillations between two states and then further bifurcations give four states. However, very tiny additional changes of r lead the way into the non-predictability of deterministic chaos. This is seen in computer simulations of equation 3.1, shown in Fig. **3.16**. The lower diagram shows the initial steady state, effective for a large range of values of r, then the first and second bifurcation leading to

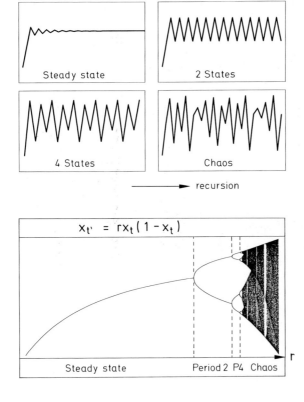

Fig. 3.16.
The route from order to chaos via increasing periods by augmenting r in the logistic recursion Eq. (**3.1**). The lower diagram shows the route from steady state via bifurcations (*period 2, period 4*) into chaos given by increasing r. The *upper four diagrams* give the calculated population sizes by iteration of Eq. (3.1) for the steady state, period 2 (two states), period 4 (four states) and chaos. (May 1976; Hastings et al. 1993)

periods of 2 and 4 states, and with increasingly smaller increments of r
there are then chaotic responses to tiny changes of r. The top four graphs
show the results of iterative calculations of population sizes x_t. At low r in
the steady state the population is stable, at period 2 there are two and at
period 4 there are 4 predictable states, while in chaos, prediction of subsequent population sizes from existing ones has become impossible.

This little excursion to population theory was necessary to explain some
basic implications of the chaos theory. Chaos is a property of non-linear
dynamic systems and these are the rule and not the exception both in the
living and non-living world. As we reject the climax theory of formation of
stable steady state vegetation types and adopt the oscillating mosaic model
with continuous dieback and renewal of "unpredictable irregularity" (see
above) for tropical forests, we are right back in the realm of deterministic
chaos.

While the theory of deterministic chaos has already had some impact in
population biology (May 1976; Hastings et al. 1993) it is intriguing that in
ecology in general it has only been accepted very reluctantly (Linsenmair
1995) or even been rejected in the exclusive distinction of deterministic and
stochastic development of diversity. It is intriguing because population biology is a field so close to or even part of ecology.

Another way of looking at the relationships of Eq. (3.1) is to consider the
control parameter r as an indicator of general resources or even as stress. If
then we envisage that a high state of order may be given by complexity,
which integrates functioning of diversity (Cramer 1993), we realize that this
must occur only within a rather narrow window of stress conditions, as it is
in fact borne out in experiments like those of Grime et al. (1987) and Tilman (1982) described in Sect. **3.3.1.2**. Thus, the chaos theory is very important in ecology (Stone and Ezrati 1996), and it is not unlikely that deterministic chaos, which certainly governs the ecology of populations, also determines the structure of tropical forests (as of other environments) with their
oscillatory and non-linear behaviour.

Although chaotic systems totally lack long term predictability their short
term predictability is better than that of stochastic random processes. One
can determine the error of such predictions, and with the appropriate algo-
rithms one can use them to regulate a system via a transputer so that the
system is forced to stay in one of the chaotic paths (or trajectories) in time
and space. Such fine regulation of chaotic systems, which offers opportunities for much more delicate manipulation than regulation of deterministic
systems, is currently assessed in physics (Hübinger et al. 1993; Schuster
1995) and even explored for practical applications in engineering. Biological systems must obey physical laws. Conversely, we may also say that
biological systems use physical laws to develop the diversity of life. It would
be most surprising if the wide scope of possibilities inherent in deterministic chaos had not been used by life during evolution. It may be noted that
with pure stochastic randomness life would be deprived of any significance

and would be cast into meaninglessness. Conversely, pure deterministic regularity, allowing one to retrace all things accurately in the past and predict them precisely in the future for all (mathematical) infinity would cast life into tedious monotony. Only deterministic chaos, with its strict mathematical rules and yet high variability, with its unpredictability and yet delicate means of fine regulation, provides an opportunity for the adaptability, plasticity, diversity and beauty of life to unfold (W. Martienssen, quoted after a public lecture; Lloyd and Lloyd 1995).

3.3.2
Vertical Structure: Strata

The vertical structure of tropical forests is determined by several more or less distinct and typical canopy layers. In simplified terms one may distinguish three major layers:

– a layer of emerging giant trees up to 60–80 m tall,
– an intermediate main canopy layer up to 24–26 m, and
– a lower canopy layer

(Whitmore 1990), as shown in the schematic transect of Fig. **3.17**. More realistic transects of actual forests often show a larger complexity, and there is also much diversity of vertical structures among forest types.

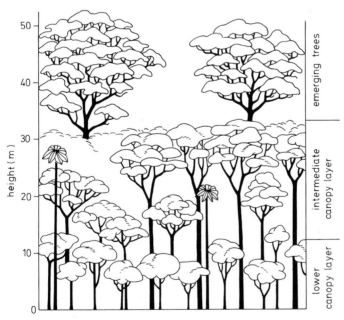

Fig. 3.17. Schematic representation of the strata structure of a tropical forest

Fig. 3.18.
Light penetration through the canopy of a tropical forest. (After Jacobs 1988)

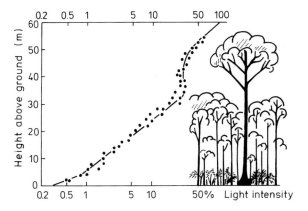

The abundant plant life in these various strata determines vertical gradients of many important environmental factors such as

– light intensity and spectral composition,
– temperature,
– air humidity,
– CO_2-concentration,
– mineral nutrients.

Due to absorption by the foliage, light intensity may decrease exponentially from the intermediate main canopy layer down to the forest floor, which often obtains only a few per cent of the intensity received by the upper canopy or by a large forest clearing (Fig. **3.18**). Light-absorption by the canopy of forests may be treated according to Lambert-Beer's law, which is well known from photometry. The ratio of the light intensity I of a beam passing through a sample solution of a thickness $\underline{d}$ and the light intensity of the incident light beam I_0 is given by

$$I/I_0 = 10^{-\varepsilon \underline{d} \underline{c}} , \tag{3.2}$$

where ε is the molar extinction coefficient and $\underline{c}$ the concentration of the sample. In analogy, the situation for canopies may be written as

$$I_{\underline{l}} / I_{\underline{o}} = e^{-\underline{k} LAI_{\underline{o} - \varepsilon}} . \tag{3.3}$$

Here I_0 is the light intensity outside the canopy and I_1 the intensity at level $\underline{l}$, $LAI_{\underline{o} - \varepsilon}$ is the leaf area index between the top of the canopy, $\underline{o}$, and level $\underline{l}$, and $\underline{k}$ is a constant. The leaf area index is dimensionless and is given by a projection of all foliage onto a certain level $\underline{l}$ or onto the ground; it thus is the ratio of

[total foliage area above a unit area at layer $\underline{l}$] : [unit area].

Light-absorption by the canopy not only reduces light intensity but also changes the **light quality** or spectral composition. In the red region of the spectrum absorption by photosynthetic pigments changes the red : far red ratio (R : FR) so that phytochrome regulated processes, which respond to this ratio are affected (Sect. **3.6.1**). In a low-land rainforest in Costa Rica R : FR was found to be 1.23 in a large clearing but only 0.42 on the forest floor (Chazdon and Fetcher 1984).

The degree to which the canopy is closed above the forest floor or leaves openings for light penetration can be determined by quantitative image analysis of photographs taken around noon with wide-angle or fish-eye lenses pointing upwards from the ground. In addition to the diffuse light filtering through the canopy foliage, the forest floor may also obtain light in the form of **light flecks**. Light flecks occur when movements of leaves in the wind or the changing angle of the sun allow direct light penetration for intermittent periods of time. These light flecks may provide up to 80 % of the total irradiation received by the forest floor, and their intensity ranges from 10 to 70 % of full sunlight. They are important for photosynthetic productivity (see Sect. **3.6.3**). Fig. **3.19** shows that for short periods lower strata in tropical forests may obtain quite high irradiance, which may at times exceed that received by strata higher up. Thus, the gradual decline of irradiance from the top of the canopy to the ground shown in Fig. **3.18** does not always correspond to the actual situation.

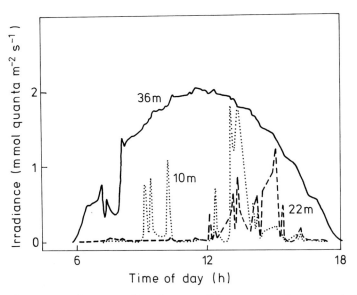

Fig. 3.19. Daily course of irradiance at different strata of a wet tropical forest in northern Australia on an almost cloudless day; i.e. *36 m* = above the canopy, *22 m* and *10 m* = increasingly lower levels inside the forest. (After Doley et al. 1987)

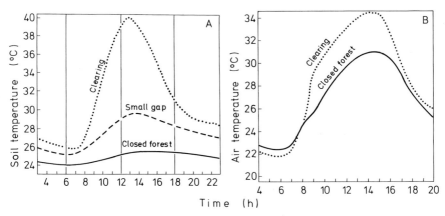

Fig. 3.20A, B. Daily course of soil temperature at 2 cm depth (**A**) and air temperature 1.5 m above the floor (**B**) in a forest in Surinam. Comparisons between the closed forest and clearings are made for the hot and dry season. (After Jacobs 1988.)

Diurnal variations of **soil temperature** and **air temperature** 1.5 m above the ground within a forest in Surinam in comparison to small and large clearings are shown in Fig. **3.20**. There is strong dampening of daily changes of temperature inside the forest, where it remains much cooler than in the gaps and clearings. **Air humidity** is related to temperature. Hence the **water-vapour pressure saturation deficit** of the atmosphere at the top of the canopy and at various levels inside the forest shows a similar pattern (Fig. **3.21**). It is close to zero at all levels at sunrise, shows a maximum at 14.00 h and decreases inside the forest from the higher to the lower strata. Since water-vapour pressure deficit of the atmosphere determines the driving force for transpiration, it constitutes a highly important ecophysiological factor.

Carbon-dioxide concentration in the atmosphere inside forests is influenced by photosynthesis and respiration of the organisms living in the

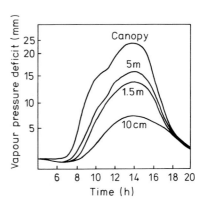

Fig. 3.21.
Daily course of water-vapour pressure saturation deficit of the atmosphere at different strata of a forest in Surinam during the dry season; i.e. above the canopy and at increasingly lower levels inside the forest as indicated. (After Jacobs 1988)

forests. Daily averages of CO_2-concentration at the soil surface may be quite large, i.e. up to 1000 ppm, due to the respiration of plant roots and soil organisms. One meter above the floor of two forests of the upper Rio Negro Basin in Venezuela, daily average CO_2-concentration was still 508 and 541

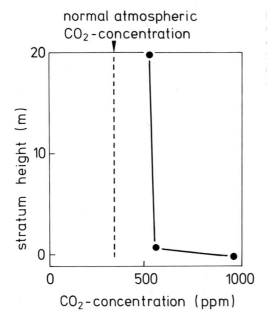

Fig. 3.22. CO_2 concentrations of the air at different levels above the soil surface in wet tropical forests. Values are daily averages. (After data of Medina et al. 1986)

Table 3.1. Mineral nutrients in rainwater, throughfall and stemflow. (A) Average nutrient concentration in rainwater and annual input in several tropical locations throughout the world. (Data from Medina and Cuevas 1994) (B) Values measured in a forest in Central-America. (Data from Junk and Furch 1985)

Nutrient	A		B		
	Nutrient concentration in rainwater (µmol/l)	Annual input via rainwater (mmol m^{-2})	Nutrient concentration (µmol/l)		
			Rainwater	Throughfall	Stemflow
Na	n.d.	n.d.	5.2	11.7	91.7
K	1.0	20	2.7	31.8	168.7
Ca	11.1	200	1.8	6.3	43.0
Mg	8.2	152	0.8	7.8	39.9
NH_4^+-N	25.9	493	12.1	< 3.6	657.1
NO_3^--N	9.3	196	7.9	40.0	19.3
PO_4-P	0.7	15	0.1	4.9	3.1
SO_4-S	21.8	424	n.d.	n.d.	n.d.

ppm respectively, and then showed little decline for up to 20 m (Fig. **3.22**). Hence, plants within the canopy may photosynthesize at CO_2-concentrations 150-200 ppm above the average CO_2-concentration in the atmosphere outside the forests (Medina et al. 1986).

Finally, even **mineral nutrients** show variability over the horizontal strata of forests, because interaction with leaves and stems causes precipitation to become enriched in nutrients. Thus the throughfall of rain and the stemflow in the forests show much higher concentrations of most mineral ions than the rain water itself (Table **3.1**).

3.4
Plant Types and Life Forms in Tropical Forests

The relationships between diversity of species and forest types has been discussed above. However, diversity is also given by the variability of **life-forms**. The life-form concept is defined rather loosely. We shall look at a few examples and show how diversity indices may be deduced.

Many attempts have been made to categorize the diversity of plant types by distinction of life forms which are morphologically more or less conspicuous. Life forms represent morphological adaptations to environments or towards a given stress factor or set of stress factors. An example illustrating an approach to such a classification is Raunkiaer's crisp distinction of five life forms of cormophytes according to the life time of shoots and the position and protection of regenerating buds, namely

- the **phanerophytes** with regenerative buds higher than 50 cm above ground,
- the **chamaephytes** with buds closer to the ground (10–50 cm above ground),
- the **hemicryptophytes** having a close contact of the buds with the ground,
- the **cryptophytes** with below ground regenerative organs (rhizomes, onions, bulbs, storage roots etc.), and
- the **therophytes** or annuals.

These life forms are particularly conceived for analysis of vegetation in mesic and temperate climates as they describe strategies for over-wintering. However, they may also be used to describe adaptation to regular seasonal drought periods and, may thus, be of some use with respect to the tropical environments. Moreover, of course, the distinction between trees (macro-phanerophytes), shrubs or bushes (microphanerophytes) and herbs of different forms inherent in these definitions is always applicable. However, this scheme is hardly sufficient to come to grasp of diversity in a tropical forest.

Vareschi (1980) has noted that as a basis for schemes defining different life forms, morphological modifications of any of the major plant organs

Fig. 3.23A-D

Fig. 3.23A-G. Root types of trees in tropical forests. Examples of buttress roots in the cloud forest of Rancho Grande, northern coastal range of Venezuela (**A, B**), of *Ceiba pentandra* (**C**) and *Pimenta racemosa* (**D**) on St. John Island (US Virgin Islands, Lesser Antilles). E-G Stilt roots of *Pandanus* (**E**) in Queensland, Australia, and a palm (**F, G**) in a forest of the Gran Sabana, Venezuela, with lenticels clearly seen in **G**

could be chosen, e.g. life forms based on roots, shoots, leaves, flowers or propagules. Another approach would be to derive life forms which use morphological modifications at the whole plant level with distinct mechanistic relations. Both shall be illustrated here by giving a few specific examples.

3.4.1
Life Forms Based on Root Categories

Root categories of trees frequent in tropical forests are stilt roots and buttress roots (Fig. **3.23**). It has been debated whether such roots have mainly mechanical functions or serve aeration and O_2-supply to below ground root tissues. Presumably both is important. Buttress roots, in particular, may function like ropes with effective anchoring of the trunk to the ground (Mattheck 1992). In addition the increase in above-ground root surface brought about by these root types may facilitate aeration. In wet tropical soils, where gas diffusion is limited and where vigorous soil-respiration will lower O_2-concentration, this may be a particularly important aspect. Indeed, immediately below ground the buttress roots show much branching and produce many fine absorptive roots, which can be supplied with O_2 via pores in the bark of the above ground buttress. Stilt roots also are often covered with lenticels facilitating gas exchange with the atmosphere (Fig. **3.23**).

3.4.2
Life Forms Based on Leaf Categories

The most variable plant organ in form is the **leaf.** Vareschi's (1980) leaf analyses of plants in the cloud forest of Rancho Grande in Venezuela reveal more than 300 forms, most of which are reproduced in Fig. **3.24** for the sake of their graphic attractiveness. **RAUNKIAER's size classes of leaves** offer a more systematic approach, where leaves are distinguished by their area:

- megaphyll $\quad\quad\quad\quad > \; 1500$ cm^2,
- macrophyll $\quad 1500 \; - \quad 180$ cm^2,
- mesophyll $\quad\;\; 180 \; - \quad\; 20$ cm^2,
- microphyll $\quad\;\;\; 20 \; - \quad\;\; 2$ cm^2,
- nanophyll $\quad\quad\;\; 2 \; - \quad 0.2$ cm^2,
- leptophyll $\quad\quad\quad\;\; < \quad 0.2$ cm^2.

Table **3.2** gives an idea of the percentage distribution of leaf sizes in a rainforest and an evergreen bushland. It shows that larger leaves predominate in the rainforest whilst smaller leaves are found in the bushland. Naturally, the leaf shape is an essential additional feature of diversity (Fig. **3.24**). It is difficult, however, to delineate distinct categories, and without any mechanistic basis this approach may devalue the use of life form classifications. Vareschi

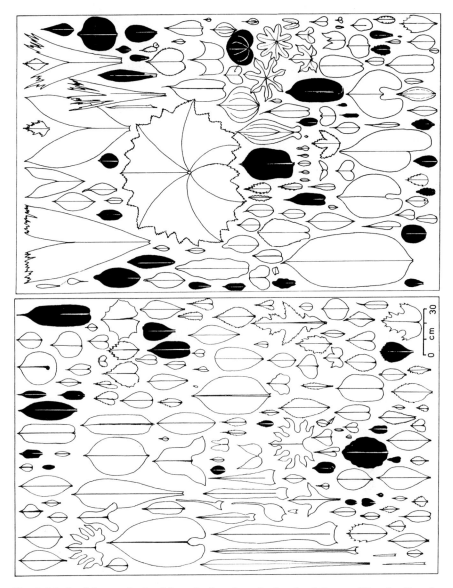

Fig. 3.24. Leaf-forms obtained from leaf analyses in the cloud forest of Rancho Grande, Venezuela (Vareschi 1980, with kind permission of R. Ulmer)

(1980) lists about 18 leaf forms according to shape and special surface features, in addition to Raunkiaer's size classes, and then derives a diversity coefficient, c_d as follows

$$c_d = n \cdot f,$$ (3.4)

where $\underline{n}$ is the number of species occurring and $\underline{f}$ the number of leaf categories. The lowest possible value of c_d is 1 (one species only occurring). In mesic environments with few species and low leaf form diversity it may be several tens or hundreds, whereas in tropical environments under favourable conditions it may reach up to 20 000.

Table 3.2. Percentage distribution of leaf sizes in an evergreen rainforest in Brazil and an evergreen bushland at Port Henderson Hill, Jamaica. (Medina 1983)

	Rainforest macrophanerophytes	Bushland microphanerophytes
Number of species:	49	43
Macrophyll	2	–
Mesophyll	75	16
Microphyll	16	74
Nanophyll	2	7
Leptophyll	4	2

3.4.3
Life Forms Based on Whole Plant Modifications with Mechanistic Relations

3.4.3.1
Epiphytes, Hemi-Epiphytes and Lianas

Very typical special life forms of cormophytes in tropical forests are epiphytes, hemi-epiphytes and lianas to which a special chapter (Chap. 4) is devoted.

3.4.3.2
Myrmecophytes

Another example of a typical life form in tropical forests are **ant plants** (**myrmecophytes**). The **symbiosis between plants and ants** plays a particularly important role in the tropical environment. It is frequently found in epiphytes (see Sect. **4.4.2**) but also in terrestrial plants, where the genera *Tococa* (Melastomataceae), *Cecropia* (Moraceae) and woody Leguminosae are the best known ant plants. In the Sira-mountains of Peruvian Amazonia Morawetz and Wallnöfer (1992) counted that 4.4 % of all species were genuine myrmecophytes. The ant plants prefer disturbed sites, which may be naturally due to land-slides; 86 % of all genuine myrmecophytes (38 species) were found in such sites, while only 14 % (6 species) regularly occurred in the primary forest.

The plants provide hollows, so-called **domatia**, where the ants find protected spaces for nests, e.g. in the inflated leaf bases of *Tococa* (Fig. **3.25A**),

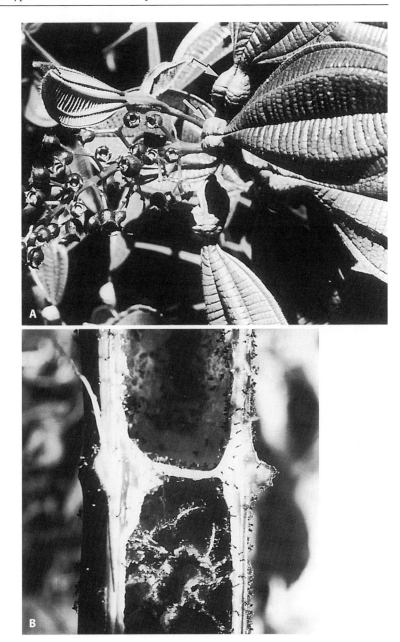

Fig. 3.25A, B. Ant-nest plants. **A** *Tococa* sp. with ant nests in the inflated leaf bases (*arrows*), **B** *Cecropia* sp. with ant nests in the hollow stem

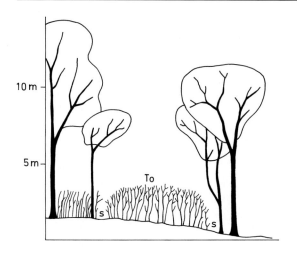

Fig. 3.26.
Stand of *Tococa occidenta-lis* (*To*) with a surrounding safety corridor (*s*) and adjacent vegetation. (After Figure 2 in Morawetz et al. 1992, with kind permission of the author and Chapman and Hall)

the hollow stems of *Cecropia* (Fig. **3.25B**) and of thorns and leaf petioles of many Leguminosae (Caesalpiniaceae, Mimosaceae). In addition the ants may receive nutrition in the form of nectar from **extra floral nectaries** or special nutritive appendices containing fat and oils, protein and carbohydrates, e.g. the **elaiosomes** (appendices of seeds) or the **Müller-bodies** of *Cecropia* which are made up of glycogen and not the usual plant-storage carbohydrate starch.

It is thought that in return the ants provide to the plants **protection from phytophagous animals** especially insects (Davidson and Epstein 1989). It also has been observed that ants keep their host plants free of epiphytes.

A most astonishing story has been reported by Morawetz et al. (1992). The *Myrmecochista* ants of *Tococa occidentalis* systematically and rapidly kill all angiosperms coming closer than 4 m to their host plants. *T. occidentalis* is a light demanding species and its growth and proliferation is stimulated by the ants' clearing of the surrounding competitors. Thus, after an initial *T. occidentalis*-plant has been colonized by the ants, pure *T. occidentalis* stands with a diameter of several meters (10–30 m) may then develop, around which the ants even maintain a "safety corridor" (Fig. **3.26**). Using their mandibles the ants cut the veins of the leaves of the competing plants. In the case of palmate leaves they attack the point at the base, where all veins join; pinnate venation is destroyed by cutting the first and second order veins at the base; the veins of moncotyledons (e.g. palms) are cut one by one along the entire leaves. After cutting the veins the ants inject a poisonous excretion from their abdomen. Apical meristems are also attacked. In this way the plants die rather rapidly. Although the ants can effectively kill 10–50 m tall trees, they do not usually attack such emergent trees at some distance of their host plants which may then form a closed canopy 10–25 m above the stand of *T. occidentalis*. As the light demanding *T.*

occidentalis-plants die away back in the shade, the stand deteriorates and the ants emigrate to start a new cycle elsewhere.

The plant-ants may also contribute to **nutrition** of their hosts (see Sect. **4.4.2**). In *Tococa guianensis* it was observed that one of the two adjacent domatia at the base of each leaf (see Fig. **3.26A**) is used for nesting and the other for dumping excrement and debris. From radioactive tracer studies the inner surface of the domatia has been shown to be absorptive of low molecular substrates like amino acids and phosphate, in contrast to the surfaces of the leaf lamina. Thus, it is highly likely that nutrients are absorbed from the rotting material in the trash-domatium (Nickol 1992).

3.5
Soils and Mineral Nutrition in Tropical Forests

It is frequently assumed that the major portion of minerals in wet tropical forests is bound in the living biomass. This is not always true and the soil may contain a considerable fraction of the minerals in the ecosystem. It is often the soil which is the most vulnerable part in tropical forests. Exposed by unbalanced logging systems or methods of shifting agriculture it may rapidly become oxidized and eroded.

An example of **inorganic nutrient cycling** in a wet tropical forest is given in Fig. **3.27**. Input of minerals to the soil is via rain, canopy leaching with throughfall and stemflow (see also Table **3.1**) and via litter fall.

Tropical forests often have a nutrient limitation of some sort. As in savannas (Sect. **7.3.2**) phosphorus is often the most problematic element and mycorrhizal symbioses between plant roots and fungi are important (Medina and Cuevas 1994). P/N-ratios in canopy leaves of tropical humid forests range from 15 to 35 (mol/mol) $\times 10^{-3}$ (Medina and Cuevas 1994) similar to those in plants of savannas, and hence are discussed in more detail in Sect. **7.3.2**. **Nitrogen cycles** for a semideciduous forest are shown in Fig. **3.28**.

Roots, which often form very dense mats, are mainly restricted to the upper 0.1 – 0.3 m of the soil, and frequently the soil layer itself on top of the bed rock is rather thin. **Soil respiration**, which is very high in wet tropical forests corresponds to 600 – 670 g organic matter m^{-2} $year^{-1}$, of which 67 – 82 % is due to respiration within root mats. **Rates of mineralization** of organic litter are high and **recirculation of minerals** is rapid.

Nutrient availability also affects the structure and longevity of leaves of forest trees. Small leathery leaves (*"scleromorphic microphylls"*) are developed on infertile soils due to N- but mainly P-deficiency (Medina and Cuevas 1989). Such leaves are more durable and better protected from herbivory (Choong et al. 1992) than large, thin leaves. Thus, nutrient investment in leaf structure provides a return in the form of photosynthetic products for a longer period of time. Deciduous and evergreen species coexist in tropical dry forests. They differ greatly in their investments of resources for

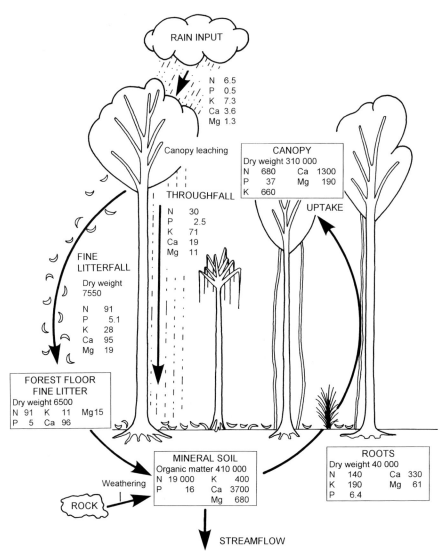

Fig. 3.27. Diagram of inorganic nutrient cycling in the lower montane rainforest at Kerigoma, New Guinea. Numbers are in kg ha^{-1} year^{-1} for flows (ar*rows*) and in kg ha^{-1} for pools (*boxes*). (After Whitmore 1990 by permission of Oxford University Press)

leaf construction and maintenance. In deciduous species, with roots occurring under relatively nutrient-rich conditions, leaves can have a potentially high nitrogen-use efficiency (CO_2-assimilation related to leaf N-content, see below Sect. **3.6.2.2**). Conversely, in evergreen species with lower nitrogen-use efficiency, the long residence time of nitrogen is favourable because roots occur in nutrient-poor soil microhabitats (Sobrado 1991). Both

Fig. 3.28.
Compartmentation and annual turnover of nitrogen in a semideciduous forest in Ghana. (From the Ecology of Neotropical Savannas by Guillermo Sarmiento. Copyright © 1984 by the President and Fellows of Harvard College. Reprinted by permission of Harvard University Press)

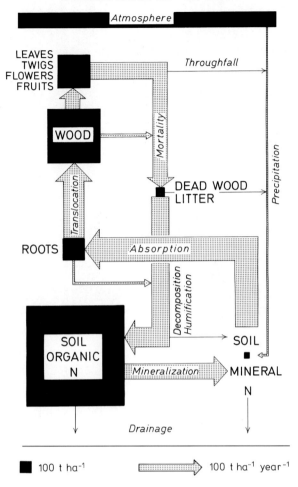

Semideciduous Forest

deciduous and evergreen species preserve nitrogen resources. Reserves of nitrogen are maintained in the twigs in drought-deciduous species and in the older leaves in evergreen species, providing some nitrogen for the reconstruction of new leaves following drought and during leaf exchange respectively (Sobrado 1995).

Different partitioning of inorganic nitrogen assimilation between the roots and shoots of trees of contrasting plant communities has been observed in South-Eastern Brazil. Rapidly growing pioneer or colonizing tree species, which are exposed to high irradiation, exhibited a large capacity to assimilate nitrate in their leaves, where light energy can be directly used in photosynthetic nitrate reduction. Leaves of shaded species had low

Fig. 3.29A, B. Leaf-flushing. A in a mango tree, B in *Brownea* sp.

levels of nitrate reductase and showed little capacity to utilize nitrate, even when it was readily available. Partitioning of NO_3^--assimilation between roots and shoots was strongly related with average daily photosynthetically active radiation rather than the availability of NO_3^- in the soil (Stewart et al. 1992). Species of closed forests often have only low nitrate reductase activities in their roots which indicates a possible preference of nitrogen uptake in the form of ammonia (Stewart et al. 1988).

In conclusion, plant species obviously allocate resources either to obtain a high photosynthetic assimilation rate from large and fragile leaves for a brief time or to provide a resistant physical structure which results in a lower rate of CO_2 assimilation over a longer time (Reich et al. 1991). Thus, mineral nutrition influences the lifespan of leaves.

Another interesting phenomenon is **leaf flushing**, which may also be related to nutrient budgets. New leaves and shoots expand from their buds very rapidly to attain a size close to that of mature leaves, much before they reach their final rigidity and pigmentation. In fact they hang down from the branches as if wilted, and often are coloured brightly yellow or red (Fig. **3.29**). The development of chloroplasts and the photosynthetic apparatus is delayed which are both particularly nitrogen-demanding. This can be considered to be an adaptation to conditions of high fungal and herbivore damage to the expanding leaves. Damage may be 100 times higher to young than to mature leaves. Mature leaves are better protected (Kursar and Coley 1992). Costs of damage to the newly flushed leaves remains low since not so many resources have been invested in them. Resource allocation to leaves becomes beneficial when they mature and establish photosynthetic productivity in return. It would be interesting to know if the bright colour of freshly flushed leaves even functions in attracting herbivores to these "cheap" leaves, thus protecting the "expensive" mature leaves. In a tropical dry-deciduous forest and a dry-thorn forest in India, phenological strategies have also been observed in relation to leaf flushing. Flushing occurs in the dry season and reaches a peak before the onset of the rains. Herbivorous insects emerge with the rains and attain a peak biomass during the wet months, so that early leaf flushing and maturation provides protection (Murali and Sukumar 1993).

3.6
Light in Tropical Forests

Due to the high leaf area index (Sect. **3.3.2**) in tropical forests, light is the dominant factor determining plant life in the forests. Light has signalling functions for the development of plants in the forest (Sect. **3.6.1**) and is the energy source for CO_2-assimilation and the formation of new organic matter or biomass (Sect. **3.6.2**), which shall be discussed separately.

3.6.1
Signalling Functions of Light: Seed Germination, Seedling Establishment and Growth

3.6.1.1
Light Quality in Tropical Forests and the Phytochrome System

As already mentioned above (Sect. **3.3.2**) light quality changes in relation to horizontal and vertical structure of forests, because filtration by canopies eliminates the red light (R) from the solar spectrum much more effectively than the far-red light (FR). Sunlight has a mean R/FR ratio of 1.2 and under green canopies the ratio may be reduced to levels below 0.5 (Vázquez-Yanes and Orozco-Segovia 1993). This affects all processes regulated by the **phytochrome system**. Irradiation with red light generates the active P_{FR} form of phytochrome, which elicits various photomorphogenetic responses. Far-red light inactivates the phytochrome, shifting the phytochrome equilibrium towards the inactive P_R-form (Box **3.2**). The light intensities required in phytochrome effects are often extremely low. It is the signalling function of light, which is sensed by the phytochrome system and not its function as an energy source.

Box 3.2

The reversible phytochrome system

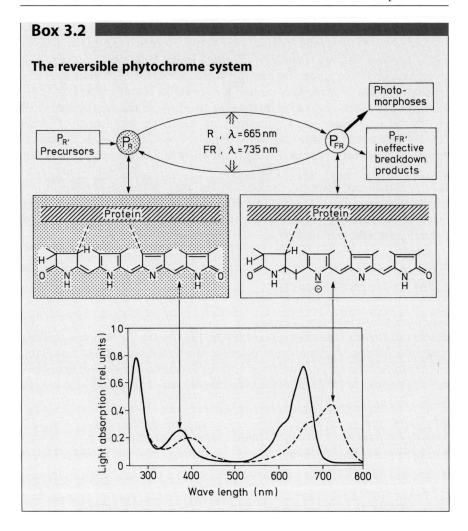

3.6.1.2
Regulation of Seed Dormancy and Germination

Among the many processes governed by phytochrome one of the most well known is the germination of the seeds of "light-germinators", i.e. **positive photoblastic seeds**. In this way phytochrome also plays an important role in the regulation of succession and regeneration in tropical forests, because light dependence of seed germination is one of the most fundamental differences between pioneer species and late successional or climax species.

Only seeds of late successional and climax species can germinate and establish seedlings under deep canopy shade. These seeds germinate very soon after dispersal and also remain alive in the soil only for a short time.

The mean ecological longevity of seeds in the tropical rainforest may be one of the shortest of any plant community (Vázquez-Yanes and Orozco-Segovia 1993). The advantage of this behaviour lies in the fact that seeds are more threatened by predators and parasites in the soil environment of the tropical rainforest with continuous moisture and high temperature, than are seedlings. Thus, the **seed banks** in rainforest soils are depleted of seeds of late succession and climax species. On the other hand, seedlings may grow extremely slowly and a persistent **nursery of small plants** is built up instead of a seed bank. Flores (1992) has studied two species of the cloud forest of the northern coastal range of Venezuela and his observations give a good idea about the actual longevity of tree seedlings after germination:

- *Aspidosperma fendleri* (Apocynaceae), an emergent species, which grows its crowns above the canopy (see Fig. **3.17**),
  - germination time   5 days,
  - longevity of cotyledons   2 months,
  - longevity of 1st leaf pair   2 years;
- *Richeria grandis* (Euphorbiaceae), a canopy species,
  - germination time   20 days,
  - longevity of cotyledons   2.5 years,
  - longevity of leaves   3 years.

The small plants remain in a state of slow growth until a canopy gap provides an opportunity for stimulation of growth (see below Sect. **3.6.1.3**). The survival of the seedlings is independent of photosynthetic parameters and largely determined by morphological characteristics which are likely to provide protection from and enhance defense against herbivores and pathogens, i. e. dense and tough leaves, a well established root system and a high wood density. Thus, seedling survival of 13 tropical tree species was found to be negatively correlated to relative growth rate (RGR), i. e. both low RGR of plants raised in the shade and high RGR of plants in the sun, and to leaf area ratio and positively correlated to root/shoot ratios and wood density (Kitajima 1994; Fig. **3.30**).

Conversely, seeds of woody pioneer species are capable of **dormancy**. They are often the most abundant components of the soil seed bank in tropical forests (Vázquez-Yanes and Orozco-Segovia 1993). Dormancy may be enforced by **hard seed coats** which are impermeable to water and oxygen and need many weeks for breakdown by weathering and microbial action. However, germination is mainly determined by light.

Light may act via temperature effects, especially via **temperature alternations**, which are required by some seeds for germination. Canopy gaps and clearings lead to greater fluctuations of soil surface temperatures due to direct insolation (see above Fig. **3.20**). Often, however, germination is regulated by light quality and the involvement of the **phytochrome** system rather than by light intensity. The reversibility of phytochrome effects may be

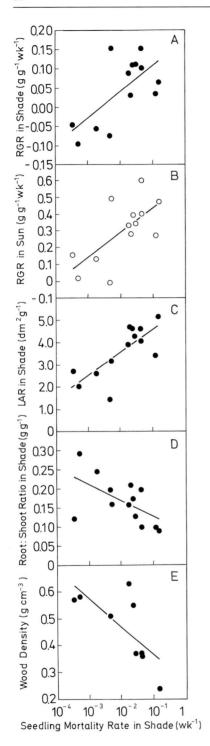

Fig. 3.30A-E.
Seedling mortality rate during the first year in the shade of 13 tropical tree species related to relative growth rate (RGR) in shade and sun (**A, B**), leaf area ratio (LAR) (**C**), root:shoot ratio (**D**) and wood density (**E**). (After Kitajima 1994)

important in excluding reactions to short **light flecks** (see below Sect. **3.6.2.3**) and sensing true light gaps. A higher proportion of red light activates and a higher proportion of far-red light inactivates phytochrome. The **photoreversibility of phytochrome mediated germination** within certain time limits may be essential to prevent germination resulting from light flecks (Vázquez-Yanes and Orozco-Segovia 1993).

The sophisticated regulation of dormancy and germination respectively, is most frequent among pioneer species and gap colonizers, with germination inhibited under closed canopies and stimulated in clearings.

3.6.1.3
Growth of Seedlings

After germination and establishment of seedlings often forming nurseries of slow growing plants (Sect. **3.6.1.2**), further growth will depend very much on **photosynthesis**, the light powered fixation of CO_2 and reduction to organic material. This, of course, applies to both pioneer and climax species. Both may differ though, in responses to light intensity, which are representative of **sun and shade plant** characteristics (see below Sect. **3.6.2**).

However, in some cases the differences in photosynthetic capacity related to light intensity between seedlings of pioneer and climax species, or between mature early and late successional species, has been found to be surprisingly small (Riddoch et al. 1991). Huber (1978) examined photosynthetic characteristics, e.g. the light compensation point (see Sect. **3.6.2.1** below), of 54 vascular plant species in Rancho Grande (Venezuela). He noted that by this criterion the majority of the species growing in the lower forest strata did not belong to extreme shade-adapted plant types, but possessed a wide capacity for response to the highly variable irradiance in this montane cloud forest. Complex regulation appears to be involved, and Strauss-Debenedetti and Bazzaz (1991) have suggested that plasticity and acclimation should be distinguished:

- **late successional species** often cannot acclimate to high light intensities when transferred from low-light to high light (**low acclimation**) but may grow well if kept continuously under low and high light respectively (**high plasticity**);
- **pioneer species** may grow at low and high light and show a considerable stimulation after transfer from low light to high light (**high plasticity and high acclimation**).

The expression of low-light and high-light forms of a species may also be determined by the **phytochrome** system (Smith et al. 1993), but in particular **blue-light photoreceptors** are also involved in this regulation (Lichtenthaler et al. 1981; Humbeck and Senger 1984; see Lüttge et al. 1986a). Leaf-anatomical features often show pronounced differences; sun leaves are

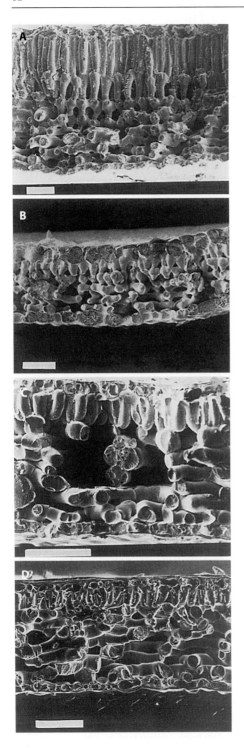

Fig. 3.31A-D.
Leaf-anatomy of seedlings of a pioneer
or early succession tree, *Nauclea
diderrichii* (**A, B**), and a late succession tree *Entandrophragma angolense*
(**C, D**), grown at high light (**A, C**) and
at low light (**B, D**). (Riddoch et al.
1991)

thicker than shade leaves and have additional layers of palisade parenchyma. In a comparison of young seedlings of the tropical trees *Nauclea
diderrichii* (De Wilde.) Merrill, a pioneer species, and *Entandrophragma
angolense* (Welw.) C.DC., a climax species, both from West Africa, differences in acclimation and photosynthetic capacity at high light intensity
were only small. However, there were marked morphogenetic effects on leaf
anatomy in plants grown in the sun and in the shade respectively, in the
pioneer species *N. diderrichii* but no so much in the climax species *E. angolense* (Fig. **3.31**).

Clearly, light also has a signalling function and does not serve only as an
energy source. We must bear this in mind, while, in the following Sects., we
return to a more classical comparison of sun and shade plants in the discussion of light-responses of photosynthesis.

3.6.2
Light Responses of Photosynthesis

Although gradients of several different environmental factors are noticeable
in tropical forests (Sect. **3.3.2**), light intensity is most highly variable and
appears to play the most prominent role in determining the ecophysiological comportment of forest plants. At the top of the canopy and in larger
clearings in full sun-light intensity of photosynthetically active radiation
(PAR) at 400–700 nm wavelength may range from well above 1000 up to
2000 μmol photons m^{-2} s^{-1}. On the forest floor there may be less 5 μmol photons m^{-2} s^{-1} (see Figs. **3.18** and **3.19**).

Thus, light can become a stress factor from both too much (when it
causes overenergization of the photosynthetic apparatus and hence photoinhibition or even photodestruction), or too little (when it becomes limiting as an energy source of photosynthesis).

3.6.2.1
Light-Response Characteristics of Sun and Shade Plants

The photosynthetic utilization of light by plants is described quantitatively by
light-response curves (Fig. **3.32**), which are distinguished by several cardinal
points as follows (see also Box **3.3**): In darkness (zero PAR), there is net-CO_2
release due to **respiration**. As light intensity increases, net-CO_2 release is
gradually reduced until the light-**compensation point** is attained, where net-
CO_2 exchange is zero because photosynthetic CO_2-uptake just balances respiratory CO_2-release. Above this point net-CO_2 uptake increases until light
saturation is reached. The **light-saturation point** often is hard to determine
precisely, because light saturation is approached gradually. Hence, **half
saturation** of photosynthesis is often quoted alternatively or additionally. The
slope of the nearly linear part of the curve below saturation gives the apparent **quantum yield** (mol CO_2 per mol photons) of photosynthesis.

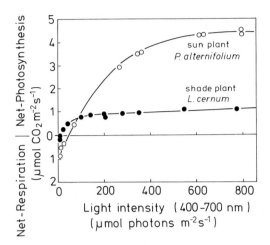

Fig. 3.32.
Light-response curves of a sun
plant *Ploiarium alternifolium* and
a shade plant *Lycopodium cernum* measured in the field in
a tropical secondary forest in
Singapore. (After Lüttge et al.
1994)

Plants may be genetically determined for growth at low or high light intensity. In this case, i.e. when plants are truely shade or light demanding, we can distinguish genuine **shade and sun species.** However, plants may also acclimate or adapt ecophysiologically to low and high irradiance, respectively (see Sect. **3.6.1.3**). Thus there may be **shade and sun forms** of given species. Plants surviving in the understory or below canopy of tropical rainforests may experience opening and closure of the canopy several times during their lifetime. Understory species of *Miconia* (Melastomataceae) responded to canopy openings by production of new sun-leaves rather than acclimation of old shade-leaves, which could significantly increase maximum assimilation rates (Newell et al. 1993). Individual plants may also have both shade- and sun-leaves when part of the foliage is shaded and exposed respectively. Then we may define such plants as shade or light tolerant, but not shade or light demanding.

All of these various aspects are very important for plant life in tropical forests with their highly variable light climates. Comparisons of light response curves and their cardinal points provide distinctive characters for shade and sun species or phenotypes (Box. **3.3**). Shade plants usually have lower rates of respiration and of photosynthesis at light saturation and hence lower light compensation and light saturation points, but higher quantum yields, than high light or sun plants. In Fig. **3.32** this is illustrated by the comparison of *Ploiarium alternifolium* and *Lycopodium cernum*, a sun and a shade plant respectively, in a secondary tropical forest in Singapore.

The different light-use characteristics of sun and shade plants are very important for understanding the distinct stages in the dynamics of tropical forests. They distinguish pioneer species from climax species, and from plants of the under-growth (Sect. **3.6.1.3**). One important difference

Box 3.3

Light-response characteristics

- *Light-response curves and their cardinal points*

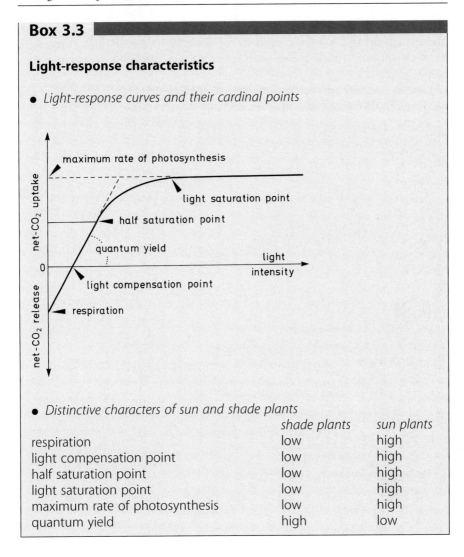

- *Distinctive characters of sun and shade plants*

	shade plants	sun plants
respiration	low	high
light compensation point	low	high
half saturation point	low	high
light saturation point	low	high
maximum rate of photosynthesis	low	high
quantum yield	high	low

between pioneer species and climax species was the light requirement for seed germination (see Sect. **3.6.1.2** above). During growth, pioneer species show the characteristics of sun plants, understory species that of shade plants, and the dominant trees of later successions (climax species) show an intermediate behaviour as they may be extremely sun exposed in the upper canopy and shaded in the lower canopy layer (Figs. **3.33** and **3.34**, Tables **3.3** and **3.4**). Thus the pioneer species *Cecropia* needs much higher light intensities for light saturation of photosynthesis and has much higher rates of maximum photosynthesis than the shade plant *Croton* (Fig. **3.33**). A comparison of the light-response curves of dominant climax trees in the upper canopy, i.e. *Cordia alliodora* and *Goethalsia meiantha,* with that of the understory shrub *Croton glabellus* in Fig. **3.33** shows the distinct differences between sun and shade plants with respect to all the features listed in Box **3.3.**

Pioneer and late successional rainforest species also regulate their leaf gas exchange in different ways. Pioneer and late stage rainforest trees have been

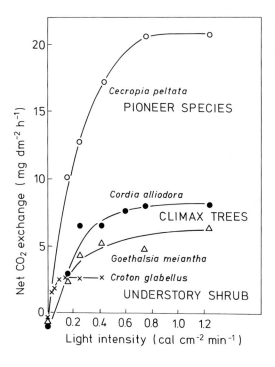

Fig. 3.33.
Light-response curves of four species of a secondary lower montane rainforest in Costa Rica (Stephens and Waggoner 1970)

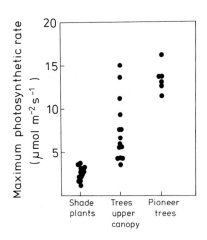

Fig. 3.34.
Maximum photosynthesis of a range of pioneer trees, upper canopy trees (climax species) and shade plants. (Medina 1986)

Table 3.3. Values for cardinal-points of light-response curves of sun and shade plants in general and of plants in tropical forests (After Lüttge 1985)

Plant type	Light-compensation point (μmol photons m^{-2} s^{-1})	Light saturation of CO_2-uptake (μmol photons m^{-2} s^{-1})	Rate of CO_2-uptake at light saturation (μmol m^{-2} s^{-1})
Sun plants	20 to 30	400 to 600	10 to 20
Shade plants	0.5 to 10	60 to 200	1 to 3
Tropical rainforest			
Upper canopy			
Sun types	12	250 to 370	13 to 19
Shade types	6 to 12	125 to 185	6 to 10
Lower canopy			
Shade types	6 to 12	125	4 to 5
Herbs of the undergrowth	2.6 to 6	25 to 37	1.3 to 1.9

Table 3.4. Maximum rates of photosynthesis of plants in various stages of succession dynamics, and daily rates of photosynthesis in various strata, of an Australian rainforest (Doley et al. 1988)

	Maximum rates (μmol CO_2 m^{-2} s^{-1})
Species of early stages of forest successions	15
Tree species of later stages of forest successions	4 – 15
Undergrowth species	1 – 3

	Daily rates (mmol CO_2 m^{-2} d^{-1})
Leaves of upper canopy level	250
Leaves at the fringe of a large clearing	97
Leaves of the undergrowth	24

studied under common conditions in artificial stands in French Guiana. It is seen that pioneer species operate at lower A/g_{H2O} ratios (CO_2-assimilation A : leaf hydraulic conductance to water vapour g_{H2O}), i.e. with a greater stomatal aperture and higher internal CO_2-partial pressures (p^i_{CO2}) as compared to trees found in the later stages of succession (Huc et al. 1994; Fig. 3.35).

3.6.2.2
The Photosynthetic Apparatus: Pigments, Enzymes and Nitrogen

It is very often observed in the tropics, that individual plants of a given species growing in deep shade inside a forest and exposed to full sun-light in an open habitat respectively, form morphologically very different **pheno-**

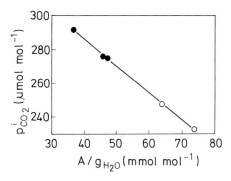

Fig. 3.35.
Leaf-internal CO_2-partial pressures ($p^i_{CO_2}$) and CO_2-assimilation (A) to leaf conductance to water vapour (g_{H_2O}) ratios of five tropical rainforest tree species in artificial stands under common conditions in French Guiana. *Closed circles*: Pioneer species (*Jacaranda copaia, Goupia glabra, Carapa guianensis*); *open circles*: late successional species (*Dicorynia guianensis, Eperna falcata*). (After data of Huc et al. 1994)

types, which are also strongly distinguished by **pigmentation.** For example, this is frequently found among rosettes of bromeliads, e.g. in the genera *Bromelia* and *Ananas* which belong to genetically identical clones propagating vegetatively by formation of ramets. In *Bromelia humilis* **shade plants** are much larger than **sun plants**; they have long and slender leaves, whereas sun plants overall have a more stunted appearance (Fig. **3.36**). The most conspicuous difference is leaf colour, which is dark green in the shade plants, brightly yellow in the sun plants and light green in intermediate forms.

What is behind this pronounced difference in pigmentation? Addressing this question requires a reminder of the basic structure of the **photosyn-**

Fig. 3.36. Phenotypes of *Bromelia humilis, from left to right* yellow/exposed form, intermediate light green/exposed form and dark green/shaded form

Box 3.4

Structure and function of the photosynthetic apparatus

A. Scheme of a chloroplast with outer membrane (envelope), inner membrane, the stroma (= plastoplasm or "cytoplasm" of the chloroplast), stromal and granal thylakoids and thylakoid interior.

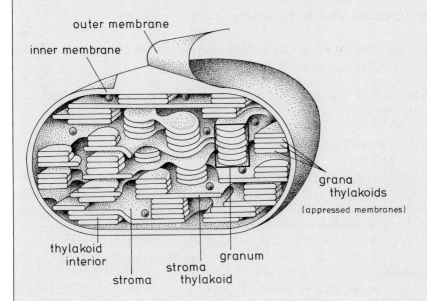

B. Scheme of the thylakoid membrane with the elements of photosynthetic electron transport

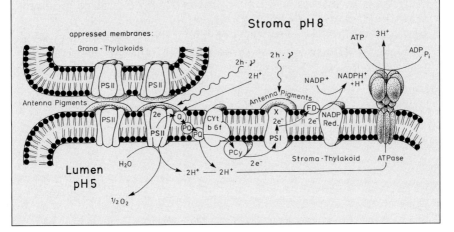

Box 3.4 (Continued)

There are four important integral proteins or protein complexes in the thylakoid membrane:

- **Photosystem II** (PS II) with antenna and reaction-centre pigments; it occurs only in the appressed regions of granal thylakoids.

- The **cytochrome b6,f-complex**.

- **Photosystem I** (PS I) with antenna and reaction-centre pigments; it occurs in the stroma thylakoids.

- The **NADP-reductase** (NADP = nicotinamide-adenine-dinucleotide-phosphate).

- The F_0F_1-**ATPase** or coupling factor.

When PS II and PS I are excited by the absorption of photons ($h\nu$), H_2O is split into oxygen, protons and electrons (e^-), electrons flow from **PS II** via membrane-bound plastoquinone (Q) and mobile plastoquinone (PQ) to the **cytochrome-b6,f-complex** and plastocyanine (PCy) at the lumen side of the membrane, and from PCy further to **PS I**, ferredoxin at the stroma side of the membrane and to the **NADP-reductase**, which finally generates the reducing equivalents needed for CO_2 reduction. If they are not utilized in CO_2 reduction, the systems of this electron-transport chain may become overreduced, and damage may result if the excitation energy cannot be dissipated in other ways.

Electron transport is associated with charge separation and simultaneously leads to establishment of a **proton-electrochemical gradient** across the thylakoid membranes between the lumen ($\sim$ pH 5) and the stroma ($\sim$ pH 8). This is the driving force for the movement of protons through the **ATPase** which is coupled to ATP synthesis providing the energy needed for CO_2 assimilation.

Box 3.4 (Continued)

C. Model of a light trap or light-harvesting complex (PS I or PS II) with antenna pigments and reaction-centre pigment.

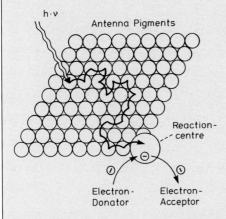

It is sufficient that one of the antenna pigments absorbs a photon (hν) and is excited. Transfer of the excitation energy between antenna pigments always eventually leads to excitation of the reaction center. The larger the light trap the more probable is excitation of the reaction centre.

Box 3.4 (Continued)

D. Photosynthetic Pigments

Chlorophyll a

System of conjugated double bonds

in Chl b

Phytyl-residue: a hydrophobic tail

Box 3.4 (Continued)

α- Carotene

System of conjugated double bonds

α - lononring

β - Carotene

β - lononring

Lutein a xanthophyll

The reaction-centre pigment chlorophyll **a** and antenna pigments chlorophyll **b** and carotenoids. The essential feature of the light absorbing pigments are systems of conjugated double bonds.

(Schemes after Lüttge et al. 1994)

thetic apparatus situated in the thylakoid membranes of the chloroplasts
(Box **3.4**). The major components are the pigments of the two photosystems
(photosystem I and II), the thylakoid proteins embedded in the membrane
lipid-bilayer and the cofactors of the photosynthetic electron-transport
chain. Pigments are the chlorophyll <u>a</u> of the light harvesting complex (light
trap) and accessory antenna pigments such as chlorophyll <u>b</u>, carotenoids
and xanthophylls. Thylakoid-proteins are the various elements of the
electron-transport chain (redox-chain), the ATP-generating coupling factor
(F_0F_1-ATPase) and chlorophyll-protein light harvesting complexes.

Since proteins as well as chlorophyll and cytochrome molecules contain
much **nitrogen**, this element plays a prominent role in constructing the
photosynthetic apparatus in the thylakoids. Given that the components of
the photosynthetic apparatus acclimate to changing light climate then
nitrogen supply is important for these processes. Thus, it is appropriate to
determine the **N-costs of thylakoid membranes**, which can be expressed in
units of mol N : mol chlorophyll. For example, in *Alocasia macrorrhiza*, a
shade tolerant species native to tropical rainforest understories in Australia,
this was 45 mol N/mol chl under natural shade, and 56 mol N/mol chl in
Pisum sativum grown at high irradiance (Evans 1988). A general comparison of shade plants and sun plants reveals a number of characteristics, as
follows:

- **Shade plants contain more chlorophyll b or have smaller chlorophyll a : b
  ratios.**

The larger relative amount of antenna pigments assures that low photosynthetically active photon flux densities (PPFDs) or light intensities are used
efficiently, i.e. at low flux densities photons are absorbed effectively and the
excitation energy can be transferred to the light trap reaction center chlorophyll (Box **3.4C**). This explains the higher quantum yield of shade plants
(Sect. **3.6.2.1**, Figs. **3.32** and **3.33**; Box **3.3**).

- **Shade plants have lower rates of electron flow along the redox-chain in
  the thylakoids related to chlorophyll.**

This is a consequence of the larger chlorophyll content of the photosystems.

- **Shade plants have less soluble protein in relation to chlorophyll.**

The soluble proteins of leaves include ribulose-bis-phosphate carboxylase/
oxygenase (RUBISCO, see also below or RUBPC in Sect. **2.5**). This enzyme-
protein is responsible for photosynthetic CO_2-fixation. It is the single major
protein and hence N-containing compound in plant leaves. The lower pro-
tein/chlorophyll ratio in shade plants is due to the higher chlorophyll content
and lower content of RUBISCO, and it is important to note, that generally

- **Shade plants have larger total N-contents in their biomass.**
- **Shade plants have larger photosystem II/photosystem I ratios.**

This is related to the change of the spectral composition of light passing through the canopy. We have already seen above (Sects. **3.3.2** and **3.6.1.1**), that the shorter-wave length red-light is filtered out to a larger extent than the longer-wave length red-light. The chlorophyll of photosystem II (PS II) is excited by somewhat shorter wavelengths (P-680 for absorption at $\lambda = 680$ nm) than that of photosystem I (PS I; P-700). Since both photosystems must co-operate in photosynthesis, shade plants need more PS II in relation to PS I.

- **Shade plants have larger chloroplasts and more grana formation.**

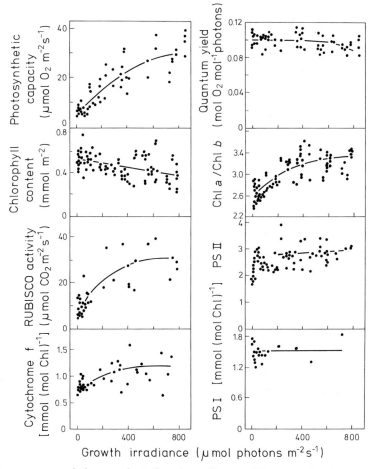

Fig. 3.37. Components and photosynthetic functions of leaves of the tropical understory plant *Alocasia macrorrhiza* grown under varying light intensities up to 800 µmol photons m^{-2} s^{-1} ($\lambda = 400-700$ nm). (After Chow et al. 1988, see Anderson and Thomson 1989)

Experiments to elucidate these relationships have often been made with tropical plants, since the contrast between deep shade in the dark rainforests and full sun exposure in clearings and open habitats with small solar inclination throughout the year is more pronounced in the tropical environment. Fig. 3.37 shows the results of a study, where *Alocasia macrorrhiza*, which has a large capacity for photosynthetic acclimation to different light environments, was adapted to various light intensities during growth, i.e. from very low PPFDs up to 800 μmol m^{-2} s^{-1}. The photosynthetic capacity, the activity of RUBISCO, the content of cytochrome f – an important element of the photosynthetic electron-transport chain (Box 3.4) – and the chlorophyll a : b ratio increased considerably with light acclimation. The amount of trap chlorophyll of PS I showed no change, but PS II increased slightly at high growth irradiance up to 800 μmol m^{-2} s^{-1}. Thus, in this particular experiment a considerably larger PS II/PS I ratio was not seen in the shade grown plants. Total chlorophyll content decreased. There was a small decrease of quantum yield. It may be noted additionally, however, that a more recent cost-benefit study which includes modelling and simulation suggests that shade leaves not necessarily have a lower photosynthetic capacity than sun leaves when leaf mass rather than area (as in Fig. 3.37) is used as a basis, although the higher investment in CO_2-fixation-cycle enzymes and electron-transport carriers per unit of leaf surface in shade plants remains evident (Sims and Pearcy 1994; Sims et al. 1994). Thus, by and large the differences between low- and high-light grown *Alocasia* plants are in conformity with the general distinctions between shade and sun plants made above, despite the general observation that *A. macrorrhiza* is a typical understory rainforest plant in Australia, and still more pronounced effects may be observed with plants showing a greater phenotypic plasticity at still higher irradiance.

A study with the low and high-light grown phenotypes of *Bromelia humilis* (see Fig. 3.36) also corroborates these basic relationships, and emphasizes the modulation by nitrogen nutrition (Fetene et al. 1990; Table 3.5). In the low-light plants, chlorophyll levels were high as compared to the high-light plants and increased by additional N-nutrition.

Table 3.5. Effects of light intensity and nitrogen on photosynthesis and leaf parameters of *Bromelia humilis* (Fetene et al. 1990)

	Irradiance during growth [μmol photons m^{-2} s^{-1}]			
	High (700–800)		Low (20–30)	
	+N	-N	+N	-N
Light-saturated rate of photosynthesis (μmol O_2 m^{-2} s^{-1})	14.0 ± 3.5	8.3 ± 2.0	16.7 ± 2.6	6.9 ± 2.5
Apparent quantum yield (mol O_2 mol^{-1} photons)	0.077 ± 0.010	0.042 ± 0.002	0.085 ± 0.015	0.041 ± 0.003
Light compensation point (μmol photons m^{-2} s^{-1})	25 ± 4	40 ± 5	10 ± 2	10 ± 5
Total chlorophyll (μg g^{-1} FW)	102 ± 21	99 ± 10	725 ± 19	364 ± 18
Chlorophyll a/b ratio	2.61	2.83	2.35	2.32
Nitrogen-use efficiency at light saturation of photosynthesis (mol CO_2 mol^{-1} leaf N)	1.2	0.5	1.0	0.4

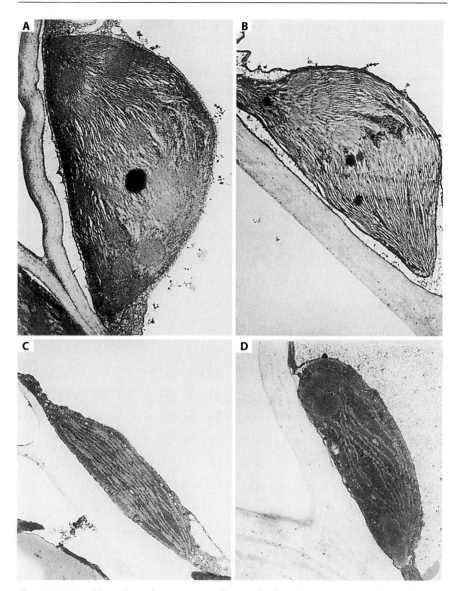

Fig. 3.38A-D. Chloroplast ultrastructure of *Bromelia humilis* grown in low light with N **A** or without N **B** and in high light with N **C** or without N **D**. (Fetene et al. 1990).

The chlorophyll a : b ratio was higher in the high-light plants. The different construction of the photosynthetic apparatus was reflected in a development of the typical cytological structure of sun- and shade-plant chloroplasts respectively (Fig. **3.38**). Low-light grown plants developed characteristically large, globular, shade-acclimated chloroplasts, with extensive grana-formation and hence appressed thylakoid membranes (Box **3.4**). The number of thylakoid membranes per granum was 3fold larger in low-light than in high-light grown plants, and the ratio of appressed to non-appressed thylakoid membranes was 3.5–5.0 and 1.0–1.5 respectively. Chloroplasts of high-light plants grown without N had only poorly developed thylakoids. Light compensation points were higher in the high-light plants of *B. humilis* (Table **3.5**). It was most noteworthy, however, that with extra N-nutrition the low-light plants could attain similar light saturated rates of photosynthesis as the high-light plants and that the high-light plants almost reached the apparent quantum yield of the low light plants. Expressed on a leaf-nitrogen basis net photosynthetic CO_2-fixation was similar in low and high-light grown plants independent of whether additional N as supplied. This demonstrates the optimisation of **nitrogen use**, but it also suggests an interaction with other factors, since the ratio of net-CO_2-fixation to nitrogen levels was not constant but rather low at low N-levels and higher at high N-levels.

Maximum assimilation rates related to nitrogen levels of leaves give the **nitrogen-use-efficiency** (NUE), which is an important parameter relating the functioning of the photosynthetic apparatus to mineral nutrition. In general, assimilation *versus* leaf-nitrogen curves are linear over certain ranges of N-levels (Fig. **3.39**). Often they do not appear to extrapolate to the

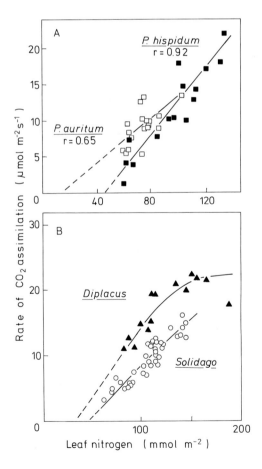

Fig. 3.39A, B.
Correlation between photosynthetic capacity and nitrogen levels in leaves of **A** *Pisum hispidum* (*closed symbols*) and *Pisum auritum* (*open symbols*), and **B** *Diplacus aurantiacus* (*closed symbols*) and *Solidago altissima* (*open symbols*). *r* correlation coefficients. (After **A** Field 1988, **B** Evans 1988)

Fig. 3.40A, B.
Photosynthesis/leaf-nitrogen relationships in
five different forest types near San Carlos de
Rio Negro, Venezuela (1° 56' N, 67° 03' W)
at ~ 100 m a.s.l. in the north central Amazon basin. Disturbed sites, high resource
regeneration sites: *a* cultivated, and *b* early
secondary successional Tierra Firme plots.
Late successional forest types: *c* Tierra
Firme, P- and Ca-limited; *d* Caatinga, N-
limited; *e* Bana, P- and N-limited. Curves
were drawn from the y-intercepts and slopes
of linear regressions given by Reich et al.
(1994) for the various samples of a total of
23 species, which they studied at the 5 sites.
The *lengths of the* lines in each case indicate
the range in which points were obtained.
Results are expressed on a mass basis **A** as
well as on an area basis **B**

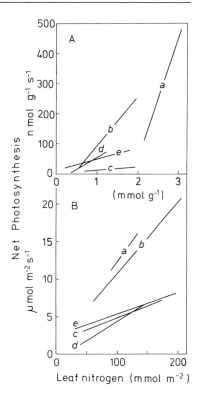

origin, and thus, show reduced NUE at low N-levels. At higher N-levels
there may be N-saturation of assimilation (see *Diplacus* in Fig. **3.39B**).
Therefore Evans (1988) has evaluated the general N-relationships of sun
and shade plants somewhat critically. While a correlation between assimilation and N-levels is clearly given, other factors like characteristics of individual species (Fig. **3.39**) and growth conditions must also be involved. This
may include irradiance during growth since with spinach and peas (unlike
in the experiment of Table **3.5**), there was an effect of light intensity on
NUE.

Reich et al. (1994) have presented a detailed investigation of these relations performed in two open and disturbed sites and three late successional
forest types in the Amazon basin, which differed in light climate and nutritional status (Fig. **3.40**). Photosynthesis-N relations are steepest and inter-
cepts on the x-axis (N-levels) are highest in the disturbed open habitats
with high resource acquisition and rapid plant growth (lines a and b in Fig.
3.40A and **B**), which also have the highest rates of photosynthesis (sun
plants). Among the other sites, under N-limitation curves were steeper
(lines d and e in Fig. **3.40A**) than when P plus Ca were more deficient (line
c in Fig. **3.40A**). This was somewhat less clearly seen, when leaf N was

expressed on a leaf area basis (Fig. **3.40B**) as compared to the leaf mass basis (Fig. **3.40A**), so that besides species and site characteristics a certain effect of the basis of data expression, i.e. mass or leaf area is also noted.

3.6.2.3
The Response to Light Flecks

Light flecks were already mentioned in relation to the vertical structure of forests (Sect. **3.3.2**) describing the dynamics of light penetrating through the forest canopy. Clearly, light flecks must be important in any type of forest. However, in the very dark, moist tropical forests they play an important role in fulfilling the energy demands of photosynthesis in lower canopy layers and particularly on the forest floor.

The dynamics of the responses of photosynthesis to light flecks in tropical plants have been studied by the group of Pearcy and Kirschbaum. A prerequisite for photosynthetic utilization of the irradiance of rapidly formed and transient light flecks are **swift and co-ordinated reactions of stomata as well as the biophysical and biochemical machineries of CO_2-assimilation.** Fig. **3.41** shows that CO_2-uptake, stomata-limited leaf conductance for water vapour (g_{H_2O}) and intercellular CO_2-concentration ($p^i_{CO_2}$) in the leaves of *Claoxylon sandwicense* and *Euphorbia forbesii* respond within minutes to a stepped increase of irradiance from low to high intensity. Both species are trees native to the understory of a mesic Hawaiian forest. In *C. sandwicense*, a tree with C_3-photosynthesis, g_{H_2O} showed an immediate linear increase, which continued for more than 50 min; net CO_2-uptake first increased very rapidly and then more gradually in correlation with g_{H_2O}; $p^i_{CO_2}$ decreased

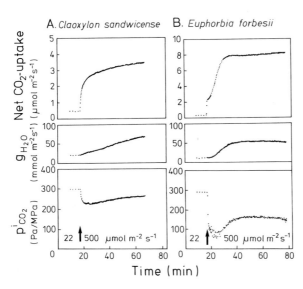

Fig. 3.41A, B. Response of net CO_2 uptake, leaf conductance for water vapour (g_{H_2O}) and intercellular CO_2 concentration ($p^i_{CO_2}$) to a stepped increase of light intensity to a PPFD of 500 μmol photons m^{-2}s^{-1} (*arrow*) after the leaves had been at 22 μmol photons m^{-2} s^{-1} for 2 h. **A** *Claoxylon sandwicense*. **B** *Euphorbia forbesii*. (Pearcy et al. 1985)

Fig. 3.42A, B.
Response of CO_2-uptake of A *C. sandwicense* and **B** *E. forbesii* during 1-min light flecks (PPFD = 510 μmol photons m^{-2} s^{-1}) on a background of 22 μmol photons m^{-2} s^{-1}, which was presented to the plants for 2 h prior to the first light fleck and which interrupted the individual light flecks. *Arrows* indicate increase (↑) and decrease (↓) of PPFD respectively. (Pearcy et al. 1985)

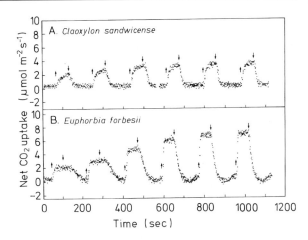

initially in response to increased availability of light energy for photosynthesis and then increased again slightly as stomata opened more widely (increased g_{H_2O}) allowing CO_2-uptake from the atmosphere. *E. forbesii*, a tree with C_4-photosynthesis (see Sect. **7.2.1**), responded in a somewhat different fashion. There was a slight delay in stomatal opening, but then maximal stomatal conductance was attained within 15 min. After an initial decline, $p^i_{CO_2}$ was stabilised at an intermediate level while CO_2-uptake reached a high, constant rate.

Are these responses sufficient to allow the plants to make efficient use of short light flecks? Interestingly there is an effect of **accelerating CO_2-uptake with time,** when very short light flecks are imposed repeatedly over short periods. Fig. **3.42** shows, for the two species discussed above, that CO_2-uptake during artificial light flecks increases gradually when subsequent light flecks of a duration of 1 min are alternated with low background intensity for about 1 1/2 min. The **intensity of this background light** is also important in maintaining the conditioning effect, which leads to increasing efficiency in the use of light flecks. As shown in Fig. **3.43,** the efficient use of 5-s light flecks at 500 μmol photons m^{-2} s^{-1} increases considerably when the background light intensity following each light fleck is increased from 0 to 10 μmol photons m^{-2} s^{-1}. Moreover, very high light intensities of light flecks are used more effectively due to the **interruptions** by low intensity background irradiation (Fig. **3.44**). When the **light fleck intensity** was below 250 μmol photons m^{-2} s^{-1}, the ratio of photosynthesis in alternating light to that during equivalent continuous illumination was < 1, but at higher intensities this ratio rose to 1.5. This suggests that there are **after-effects** on photosynthesis during background illumination, by which an intrinsic potential for reduction built up during absorption of light at high intensity can be used more effectively when the high irradiance is not continuous.

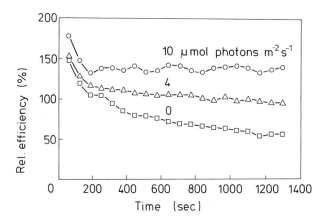

Fig. 3.43. Efficiency of the utilization of 5-s light flecks (PPFD = 500 μmol photons m^{-2} s^{-1}) on a background irradiance of PPFD varied between 0 and 10 μmol photons m^{-2} s^{-1} given for 60 s after each light fleck. In each case, leaves had been conditioned to reach a steady-state CO_2 assimilation at 500 μmol photons m^{-2} s^{-1} before applying the series of light flecks (Kirschbaum and Pearcy 1988b)

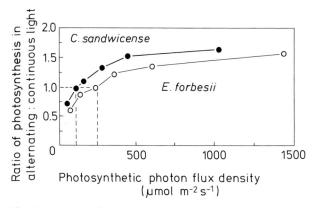

Fig. 3.44. PPFD dependence of the ratio of the apparent photosynthetic rates in alternating light to the mean of the photosynthetic rates measured during continuous illumination at the respective intensities in *C. sandwicense* and *E. forbesii*. For the alternating light, background irradiation was 12 μmol photons m^{-2} s^{-1} given for 5 s and 5-s light flecks had increasing intensities of up to almost 1500 μmol photons m^{-2} s^{-1}, as shown on the abscissa. (Pearcy et al. 1985)

What is the nature of these **conditioning processes**, which include induction, the use of short-time high irradiance and the apparent after-effects during low background irradiation? The experiment of Fig. **3.45** offers deeper insights. The tropical shade-plant *Alocasia macrorrhiza* was used in a 20-s light fleck experiment, while O_2-evolution and CO_2-uptake were recorded simultaneously.

Fig. 3.45.
Time course of photosynthetic O_2 evolution and CO_2 uptake of a fully induced leaf of *Alocasia macrorrhiza* during a 20-s light fleck (PPFD = 500 µmol photons m^{-2} s^{-1}). *Arrows* indicate duration of light fleck (Kirschbaum and Pearcy 1988a)

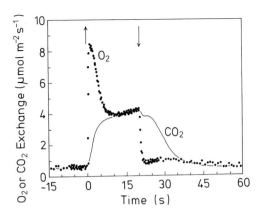

At the beginning of the light fleck, O_2 evolution increased very rapidly, and during the first second it attained about twice the rate of steady state CO_2-uptake. Subsequently it dropped again and matched the rate of CO_2-uptake after 2 1/2 s. This suggests that light-dependent electron transport, indicated by photosynthetic O_2 evolution, may proceed rapidly to fill up pools of reduced compounds in a very short initial period after stepped increase in irradiance, before it becomes limited by reactions of CO_2-reduction. Biophysical light-reactions of photosynthesis are extremely fast (Box **3.4B**). Therefore, in light flecks the slower processes of induction are the biochemical reactions of CO_2-fixation and assimilation as well as stomatal responses. Among them, the regeneration of ribulose-bis-phosphate the CO_2-acceptor in photosynthesis, is comparatively fast. It was found to be 60–120 s under transient light conditions as typical for light flecks, while light-activation of the activity of the carboxylase (RUBISCO) and stomatal reactions occurred in the range of 10–30 min (Sassenrath-Cole and Pearcy 1992).

On the other hand, the mechanisms which show slower induction are also subject to slower decay and may remain active during intermittant light flecks. At the end of a light fleck, when there is a stepped decrease of irradiance, O_2-evolution drops immediately as photosynthetic electron transport stops, whereas CO_2-uptake only declines gradually. The after-effect of high irradiance, seen in Fig. **3.45**, must be due to a surplus of reduced compounds formed during high PPFD.

This phenomenon explains the particular efficiency of short duration light fleck utilisation as **co-ordination of more rapid responses (photosynthetic electron transport) and more sluggish processes (photosynthetic CO_2 assimilation)**, with both adjusted to some extent in series so that each process may not be entirely simultaneous. Of course, the light flecks must be short, if this is to be relatively important quantitatively. Indeed, for 1-s light flecks, the integrated net CO_2-uptake (sum of net uptake in low irradiance plus integrated net uptake as a result of light flecks) is considerably

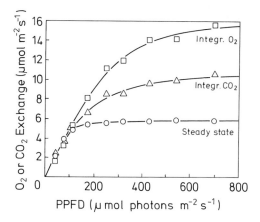

Fig. 3.46.
Photosynthetic photon flux
density (*PPFD*) dependence of
steady-state net CO_2 uptake and
integrated CO_2 uptake and O_2
evolution in *Alocasia macrorrhiza* during 1-s light flecks. For
integrated O_2 and CO_2-uptake
data give the sum of exchange
rates at low background PPFD
plus the exchange as a result of
high PPFD during light flecks
(Kirschbaum and Pearcy 1988a)

larger than the steady-state assimilation rate (Fig. **3.46**). The different time
constants of the processes involved also explain the conditioning to intermittant light. Thus, the specific dynamics of transients after light intensity
is stepped up and down, make light flecks a quantitatively more important
energy source for forest-floor photosynthetic carbon assimilation than one
might expect from their intensity and duration alone.

One may also ask whether light fleck responses are different in shade and
sun plants, since one might expect that the latter are less dependent on
intermittant light. Indeed, such differences have been observed. They are
largely based on stomatal dynamics. In species of *Piper*, acclimation of stomatal responses to different light intensities was observed, which was
important for the performance of the plants in varying light environments
(Tinoco-Ojanguren and Pearcy 1992). A comparison of *Piper auritum*, a
pioneer tree, and *Piper aequale*, a shade tolerant shrub of Mexican tropical
forests, showed that differences in induction of photosynthesis could be
accounted for by differences in stomatal behaviour. The shade tolerant
shrub, *P. aequale*, had the larger and more rapid response of stomatal conductance (g_{H2O}) to light flecks, which was shown to improve carbon gain
during subsequent light flecks for shade adapted plants. Conversely, low-
light acclimated plants of the pioneer tree, *P. auritum*, showed even slower
and smaller conductance responses than sun-acclimated plants, and there
was no significant improvement in use of subsequent light flecks (Tinoco-
Ojanguren and Pearcy 1993a,b). Another comparison was provided by
Poorter and Oberbauer (1993), who studied saplings of a climax tree species, *Dipteryx panamensis*, and a pioneer tree, *Cecropia obtusifolia*, in a
rainforest of Costa Rica. The results of their comparative investigation are
compiled in Table **3.6**. Remembering the differences in general photosynthesis characteristics of these groups of plants (Sect. **3.6.2.1**) it appears that
the climax-tree saplings exploit temporal variation in light availability by

Table 3.6. Comparison of the responses of saplings of two Costa Rican rainforest tree species to light flecks in situ (Poorter and Oberbauer 1993)

Character	*Dipteryx panamensis* Climax species in bright microsites	*Cecropia obtusifolia* Pioneer species
Induction time needed to reach 90% of light-saturated rate of photosynthesis in the morning	16 min	10 min
Daily average induction time needed	Shorter than in *C. o.*	Longer than in *D. p.*
Duration of maintenance of high levels of induction	Longer than in *C. o.*	Shorter than in *D. p.*
Behaviour when grown in shaded sites as compared to bright sites	Faster rates of induction	No difference in rates of induction
	No difference in light-saturated rates of photosynthesis	Lower light-saturated rates of photosynthesis

refining the speed of the induction response. In contrast, the pioneer species adjust by realising higher rates of light-saturated photosynthesis under high irradiation.

Different light fleck responses have also been reported in relation to leaf-longevity (Kursar and Coley 1993). In shade-tolerant species with short lived leaves (1 year) induction to attain 90% of maximum photosynthetic rates took 3–6 min, while 11–36 min were needed in long-lived leaves (> 4 years). In this case, however, RUBISCO activation seemed to be the time-limiting factor.

3.6.2.4
The Response to High-Irradiance Stress

The biological stress concept has shown us that stress can result from low or high dosage of any particular environmental factor (Box **3.1**). Shade plants, or phenotypes considered in Sects. **3.6.2.1** and **3.6.2.2**, are adapted to low irradiance stress typical of the interior of dense forests. Stress by high irradiance might generally appear to be more characteristic of open habitats like savannas. However, it is also found in deciduous and semi-deciduous dry forests and in the upper canopy of wet forests. Moreover, in view of the very high light intensities of some light flecks, it may even be a particular problem for the shade-adapted plants in the understory of moist forests. Finally, shade plants of the forest floor may be suddenly exposed to high irradiance when gaps are created by falling trees. Hence, it is necessary to address of how plants avoid damage from excess irradiation.

Box 3.5

Light absorption, chlorophyll excitation and relaxation

A. Absorption spectra of the reaction-centre pigment chlorophyll **a** and the antenna pigments chlorophyll **b** and carotenoids (see Box 3.4).

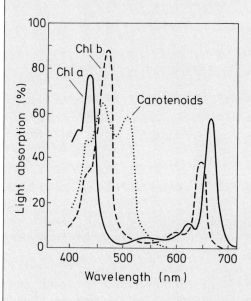

B. Ground state and excitation states with substates (*horizontal lines*) and relaxation of chlorophyll:

Absorption of the more energy-rich blue-light quanta (shorter wavelengths) leads to the **second and third excited singlet state** (half-life 10^{-14}-10^{-15} s). Absorption of the less energy-rich red-light quanta (longer wavelengths) leads to the **first excited singlet state** (half-life 10^{-9}-10^{-11} s). Relaxation can occur by transition between systems, and energy is dissipated as **heat**. Transition from the first singlet state to the **triplet state** (half-life 10^{-4}-10^{-2} s) is only probable when the whole system is overexcited. Relaxation from the first singlet state to the ground state in addition to energy dissipation as heat can occur by emission of light as **fluorescence**. Relaxation from the triplet state to the ground state is possible by the emission of light as **phosphorescence**. Energy transfer from the first singlet state and the triplet state can lead to **photochemical work**, i.e. CO_2-assimilation or photorespiration in the case of the first singlet state and formation of oxygen radicals and **photodamage** in the case of the triplet state.

Box 3.5 (Continued)

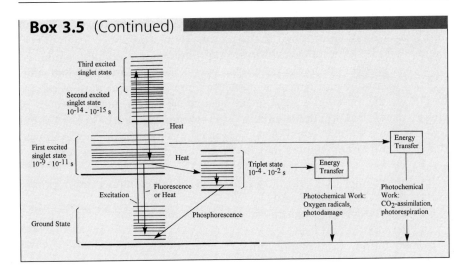

Box 3.6

Possible ways of relaxation of photosystem II

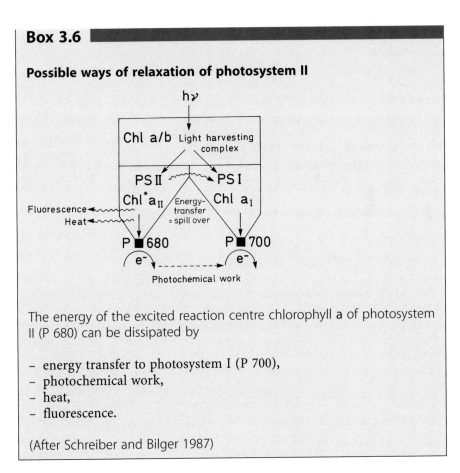

The energy of the excited reaction centre chlorophyll a of photosystem II (P 680) can be dissipated by

- energy transfer to photosystem I (P 700),
- photochemical work,
- heat,
- fluorescence.

(After Schreiber and Bilger 1987)

In this context first it is necessary to recall the various states of **excitation and relaxation of chlorophyll** in the photosystems (Box. **3.5**). Excitation to the 2nd singlet level follows absorption of photons of blue light, while red light causes excitation to the 1st singlet level. Relaxation from the 2nd singlet state occurs by emission of heat, while there are several ways of relaxation from the 1st singlet state to the ground level. The normal means of energy dissipation in photosynthesis is **photochemical work** (Boxes **3.5** and **3.6**), i.e. the eventual reduction of CO_2 fixed via RUBISCO. This, however, under certain circumstances may become a limiting process, e.g. due to

– low intercellular CO_2-concentrations in leaves following oversaturation of the photosynthetic apparatus by PPFD,
– closure of stomata to reduce transpirational loss of water in response to high irradiance and heat,
– overexcitation of the photosynthetic apparatus and overreduction of the redox-elements of the photosynthetic electron-transport chain (see Box **3.4**).

One valve for dissipation of surplus energy is photochemical work via **photorespiration**. This is possible, since RUBISCO (ribulosebisphosphate carboxylase/**oxygenase**) not only reacts with CO_2 but also with O_2 and can oxygenate ribulosebisphosphate to form phosphoglycolate, which is metabolized in the photorespiratory reaction cycle. However, this still may be of limited capacity to avoid adverse effects of overenergization, which leads to photoinhibition. The primary site of photoinhibition is photosystem II. Thus, it is important that there are several other means of dissipating excitation of chlorophyll $\underline{a}$ in photosystem II (see Boxes **3.5** and **3.6**), i.e. in addition to

– PS II **photochemistry**, due to CO_2 or O_2 binding,
  there are
– relaxation by emission as **heat**,
– relaxation by emission as **fluorescence**,
  and
– relaxation by **energy transfer** to photosystem I (Box. **3.7**),
  all of which can contribute to prevent
– transfer of excitation from the 1st singlet state to the **triplet state**, which is correlated with a change of electron spin from antiparallel to parallel (Box **3.5**).

The triplet state is much more stable, i.e. it has a much longer half-life, than the two singlet states. Therefore, it may lead to the formation of oxygen radicals with the subsequent destruction of pigments, lipids and membranes.

In summary, overenergization by high-irradiance **stress** may cause **reversible or irreversible photoinhibition** of CO_2-assimilation, representing

Box 3.7

Short-term regulation due to transfer of components of photosystem II (PS II) from the appressed grana-thylakoid regions to the non-appressed stroma-thylakoid regions in response to stress

A. Excess **light** on PS II:

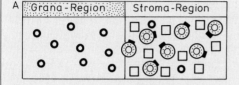

The peripheral part of the light-harvesting complex (LHC) of PS II is transferred to the stroma region, and this is dependent on its phosphorylation.

B. **Heat** stress:

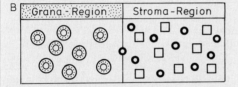

The core of the light-harvesting complex (LHC) of PS II is transferred to the stroma region.

⊚ peripheral LHC II   ⊚ peripheral phospho-LHC II

O core PS II - inner LHC II   ☐ core PS I - LHC I

(After Anderson and Thomson 1989)

---

elastic and plastic strain respectively, in the terminology of the stress-concept (Box **3.1**). During this period of strain other mechanisms of energy dissipation become dominant, and then **irreversible photodamage** may occur. In general, photoinhibition itself may be due to mechanisms of both photoprotection and photodamage, and is generally defined as representing both reversible, competing dissipatory reactions as well as the need for repair of specific system components.

Box 3.8

Fluorescence analysis

A. Fluorescence induction kinetics (Kautsky effect)

In the first second after excitation fluorescence of photosystem II is induced and rises from the fluorescence (O) of the dark-adapted leaf obtained at very weak excitation energy to I, where all primary electron acceptors (Q in **Box 3.4B**) are reduced, and after a small shoulder further to P, where the plastoquinone pool (PQ in **Box 3.4B**) is also reduced.

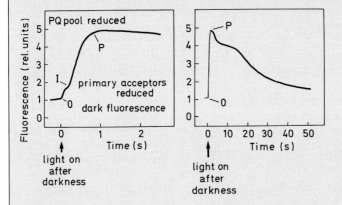

The high fluorescence obtained at P after the first second is quenched as the electron-transport capacity at the acceptor side of photosystem I and other fluorescence-quenching processes are activated.

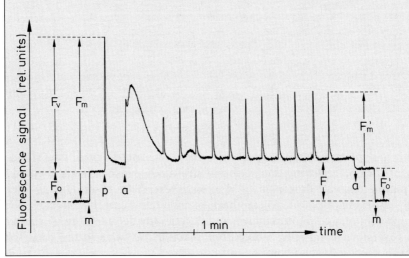

Box 3.8 (Continued)

B. Fluorescence analysis

In a pulse modulated-fluorescence-analysis system kinetics (see graph on page 110 bottom) are obtained during induction after a period of darkness (the example given was obtained from a leaf of *Clusia multiflora*):

Arrowheads pointing upwards and downwards respectively, indicate switching on and off of light, namely
- m   weak measuring light;
- p   a single pulse of saturating actinic light;
- a   actinic light with regular light-saturating pulses.

Symbols in the graph have the following meaning:
- F_o minimal fluorescence yield of dark-adapted sample in weak measuring light;
- F_m maximum fluorescence yield of the dark-adapted sample;
- F_v maximum variable fluorescence;
- F_o' minimal fluorescence yield of the light-adapted sample;
- F_m' maximum fluorescence yield of the light-adapted sample.

Calculations which can be made include the following:
1. F_v/F_m as a measure of potential quantum yield of photosystem II after dark adaptation, which is lowered by photoinhibition, so that F_v/F_m also is
$$F_v/F_m = (F_m - F_o)/F_m. \tag{a}$$

2. $\Delta F/F_m'$ as a measure of effective quantum yield,
$$\Delta F/F_m' = (F_m' - F)/F_m'. \tag{b}$$

3. $(\Delta F/F_m') \times PPFD$ as an empirical approximation of the relative electron transport rates, where PPFD is incident photosynthetic photon flux density.

4. The quenching coefficient for photochemical quenching of fluorescence, q_P
$$q_P = (F_m' - F)/(F_m' - F_o'). \tag{c}$$

5. The quenching coefficient for non-photochemical quenching of fluorescence, q_N
$$q_N = 1 - (F_m' - F_o')/(F_m - F_o). \tag{d}$$

6. The extent of non-photochemical quenching, NPQ,
$$NPQ = (F_m - F_m')/F_m'. \tag{e}$$

(After Schreiber and Bilger 1993; see also section **8.2.2**, Figures **8.17** and **8.18**.)

Among the reactions of energy-dissipation as an alternative to photo-chemical work, emission of **fluorescence** is most readily measured by photometric techniques, where the yield and **quenching** respectively, **of fluorescence** is analyzed. Fluorescence represents a competing reaction with photochemistry and with other non-photochemical processes, and hence the magnitude of fluorescence quenching quantitatively represents:

- photochemical work, i.e. transfer of electrons from excited photosystem II to plastoquinone ("**photochemical quenching**", q_P);
- energization of thylakoid membranes (i.e. "**non-photochemical quenching**", q_N);
- reversible and irreversible **photoinhibition.**

Photochemical quenching indicates effective functioning of the mechanism leading to photosynthetic CO_2-assimilation. This has led to the develop-

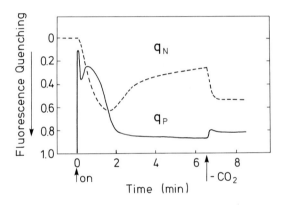

Fig. 3.47.
Time course of photochemical (qp) and non-photochemical (qN) quenching in a *Vicia faba* leaf following a saturating light pulse of 2000 µmol photons m^{-2} s^{-1} (↑ on) in normal air and after the removal of CO_2 (↑ -CO_2). (Schreiber and Bilger 1987)

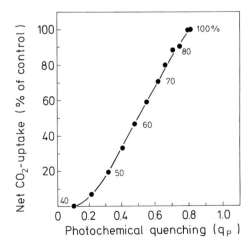

Fig. 3.48.
Correlation between CO_2 assimilation, photo-chemical-quenching (qp) and drought stress (relative water content values in % along the curve) in leaves of *Arbutus unedo*. (Schreiber and Bilger 1987)

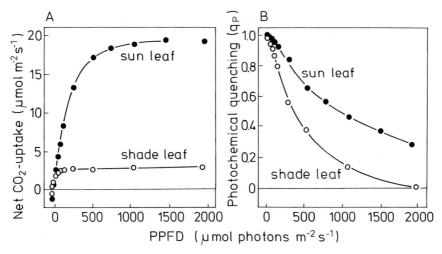

Fig. 3.49A, B. Light dependence curves of net CO_2 uptake (**A**) and photochemical quenching (**B**) of sun and shade leaves of *Arbutus unedo*. (Schreiber and Bilger 1987)

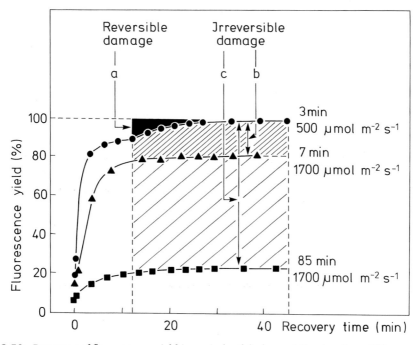

Fig. 3.50. Recovery of fluorescence yield in periods of darkness following three different photoinhibitory treatments of *Arbutus unedo*, which are indicated at the *right end of the curves* where the duration and light intensity of photoinhibitory irradiation applied to the leaves is given. Fluorescence yield is plotted in relative units, as percent of that of the non-stressed leaves. Reversible damage (*a*) and smaller (*b*) and larger (*c*) irreversible damage are also indicated. (Schreiber and Bilger 1987)

ment of sophisticated analyses of fluorescence signals, which allow to distinguish between photochemical competence and the balance between
reversible and irreversible photoinhibition, as explained in Box. **3.8.** The
method has become a powerful tool in plant-ecophysiology and is illustrated by a few examples below (Figs. **3.47 – 3.50**). It is an essentially non-
invasive and a non-destructive method, which is also potentially suitable for
remote sensing, and hence, for the ecophysiological survey of large tropical
forest and savanna areas (see Chap. **2**).

Fig. **3.47** shows the effect of light and then removal of CO_2 from the atmosphere of a pho-
tosynthesizing leaf. Photochemical quenching (q_P) is reduced and non-photochemical
quenching (q_N) is increased. However, a significant level of q_P is still maintained due to
the presence of oxygen and photorespiration. If O_2 were also absent q_P would become
close to zero. Fig. **3.48** depicts a correlation between CO_2-assimilation, photochemical
quenching and drought stress with a decrease of CO_2 uptake and q_P as leaf relative water
content decreases from 100 % to 40 %. In Fig. **3.49** we see a comparison of light-response
curves of net CO_2-uptake and photochemical quenching of a sun and a shade leaf. The
CO_2-uptake curves show the typical characteristics of sun and shade types (see Fig. **3.32**).
Photochemical quenching is high in both cases at low PPFD and decreases much more
rapidly with increasing light intensity in the shade leaf than in the sun leaf, suggesting
increased overreduction. Lovelock et al. (1994) noted that although late stage succession
species showed a greater degree of photoinhibition in forest gaps at midday than pioneer
species, it was unlikely that photoinhibition, alone, is responsible for seedling fatalities
after rainforest disturbance. Additional environmental stress, e.g. water deficits, must be
involved. Finally, Fig. **3.50** gives an example for the detection of reversible and irreversible photoinhibition caused by irradiation with photoinhibitory light intensities. Fluorescence yield was reduced due to damage by various degrees of photoinhibitory treatment.
After 3 min at 500 μmol photons m^{-2} s^{-1} it recovered during a subsequent period in darkness within about 10 min to almost 90 % of the initial level and was subsequently fully
restored within 20 min. Thus, there was a small reversible damage. After 7 min at 1700
μmol photons m^{-2} s^{-1} and 40 min recovery there remained a reduction of fluorescence
yield of about 20 %, i.e. there was a small irreversible damage. After 85 min at 1700 μmol
photons m^{-2} s^{-1} and more than 40 min recovery fluorescence yield remained reduced to
25 % of the initial value. Hence, there was a large irreversible damage although it may be
noted that recovery over longer periods may be supported by weak light which was not
studied in these experiments.

Processes of energy dissipation contribute to stress tolerance or resistance
of high irradiation. One particular mechanism, which has recently received
much attention, is based on special pigments belonging to the group of xanthophylls (Box **3.4**). The **xanthophyll-cycle,** may either be involved directly
in dissipation of energy of the singlet excited state of chlorophyll or function in determining the structure of the light-harvesting complexes (Box
3.9). The xanthophyll-cycle, for example, obviously is operative in the light-
exposed phenotypes of bromeliads in the tropics. Zeaxanthin was only
detected in the yellow high-light plants of *Bromelia humilis* (see Fig. **3.36**)
and not in the shade plants (Fetene et al. 1990) and also plays a significant
role in *Guzmania monostachia* (Maxwell et al. 1994, 1995). A survey of several other sun and shade plants including tropical rainforest species also
showed that sun plants possessed larger xanthophyll-cycle pools and greater

Box 3.9

Xanthophyll-cycle

The scheme originally proposed by Hager (1980) suggests the possible turnover in the cycle for detoxification of singlet activated oxygen and energy dissipation, namely binding of activated oxygen by oxidation of zeaxanthin to violaxanthin (epoxidation) and dissipation of the energy of photosynthetic electron transport by rereduction (deepoxidation) to zeaxanthin. The pH optima of the epoxidase (pH 7.5) and the deepoxidase (pH 5.2) correspond to the prevailing conditions in the chloroplast stroma and the thylakoid interior respectively. More recently evidence for the appropriate localization of the epoxidase on the stroma side and the deepoxidase on the thylakoid-lumen side of the thylakoid membrane has been obtained. The deepoxidase is mobile within the thylakoid lumen at neutral pH but becomes membrane-bound when the pH drops and a pH gradient is established across the thylakoid membrane by photosynthetic electron flow (Hager and Holocher 1994); it has a narrow pH optimum at pH 5.2 and is presumed to be activated by the photosynthetic electron transport via acidification of the lumenal pH. (Büch et al. 1994).

Currently, there are two contrasting schools of thought about the detailed functioning of the xanthophyll-cycle in photo-protection. One of them assumes that zeaxanthin can directly accept electrons and quench the singlet excited state of chlorophyll leading to non-photochemical energy dissipation (Demmig-Adams 1990; Demmig-Adams and Adams 1992; Pfündel and Bilger 1994). Alternatively, or in addition, xanthophylls may be involved in determining the structure of the light-harvesting complex itself (Horton et al. 1994). The pH gradient across the thylakoid membrane (Box **3.4B**) may have a function in causing conformational changes of the system, possibly involving aggregation of the light-harvesting complex of photosystem II, which may be facilitated by an absence of violaxanthin and/or a presence of zeaxanthin (Bilger and Björkman 1994). Three possible xanthophyll mechanisms are listed by Schindler and Lichtenthaler (1996), namely

- reaction of zeaxanthin with highly reactive oxygen species forming violaxanthin in a non-enzymic way;

- reaction of zeaxanthin with reactive oxygen species in an indirect way removing epoxy groups from fatty acids, where violaxanthin formation provides protection against lipid peroxidation;

- aggregation/dissociation of the light-harvesting complex (see Horton et al. 1994).

Box 3.9 (Continued)

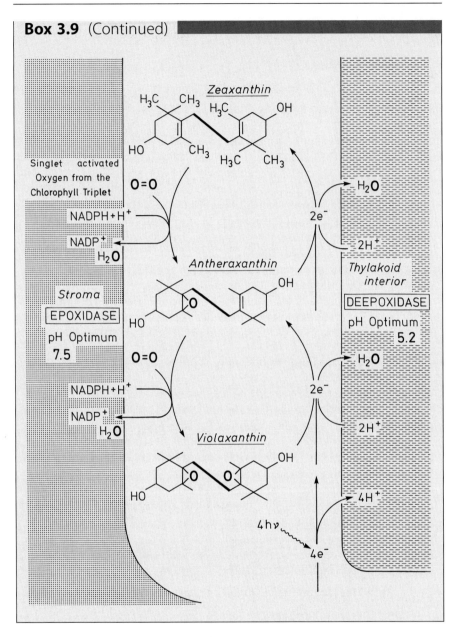

maximal zeaxanthin and antheraxanthin contents than shade plants. Sun plants displayed a greater maximal capacity for photoprotective energy dissipation via the pigments than plants acclimated to very low irradiance, and in sun leaves the reduction state of photosystem II at full sun light remained at a much lower level than in shade leaves (Demmig-Adams and Adams 1994).

Box 3.10

Damage and repair of the reaction centres of photosystem II in light-stress and recovery cycles

Under high light intensities (HL) the D_1-protein is first modified reversibly and recovery needs low light intensities (*LL*); (①). Under further stress the D_1-protein is damaged irreversibly ((②)) and repair needs protein synthesis ((③)). The turnover of the D_2-protein is slower although it also can be damaged irreversibly by light stress. (Schäfer and Schmid 1993; Critchley and Russel 1994)

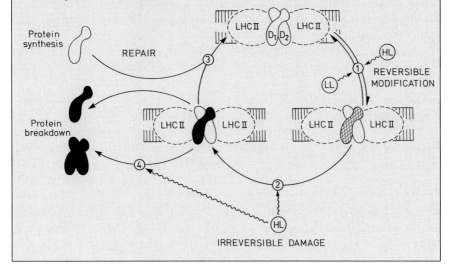

Turnover in the xanthophyll-cycle can be very rapid. Alternatively and in the absence of xanthophyll-cycle associated energy dissipation a slower process develops, which is irreversible when protein biosynthesis is inhibited. It is probably due to a functional dissociation between the reaction centers and antennae of photosystem II (see Box **3.4**; Demmig-Adams and Adams 1993). The reaction centers of photosystem II are built up of one copy each of a D_1- and a D_2-protein. The destruction of these proteins by high irradiation prevents coupling of excitation and electron transport via the light trap reaction centers. Repair needs protein synthesis (Box **3.10**). Thus, "inhibition" and "damage" may be part of protective mechanisms in that D_1-, D_2-protein damage and turnover are involved in photoprotection.

In general, we realize difficulties and ambiguities in our nomenclature: What is "inhibition" and what is "damage", what is a reaction for "protection" and what is "destruction"? The differences between acute and chronic damage may be gradual and the quantitative relationships between various ways of energy dissipation in relation to dynamics of photon-protective forces remain a conundrum (Osmond and Grace 1995).

3.7
Thornbush-Succulent Forests

The thornbush-succulent forests are separated in this final Sect. on tropical forests because they are characterized by a combination of high irradiance, low water availability and high temperature which causes a typical stress syndrome which is not encountered in this particular way in the other evergreen, seasonal or semi-evergreen forest types. An exception may be the epiphytic canopy habitat but this is covered in a separate chapter (Chap. 4). In response to this stress there is not only modulation of gas exchange of C_3-plants but a distinctive modification of photosynthetic metabolism namely crassulacean acid metabolism (CAM), seems to be a particularly useful trait in many species encountered among the plants of thornbush forests.

3.7.1
Physiognomy

Cactus-forests represent an example of thornbush-succulent forests, and have already been mentioned above in Sect. 3.2. This type of forest in Madagascar covers a small area in the south-west of the island (see Fig. 2.13) dominated by Didieraceae (11 species in 4 genera), Euphorbiaceae and other succulent and deciduous woody species (Fig. 3.51). It is also characteristic of the "Caatinga" formation in NE-Brazil (Fig. 3.52A). In Venezuela there is the "Espinar" in the area around Carora, where Vareschi (1980) distinguishes a thornbush-forest, with Mimosaceae (*Haematoxylon praecox*) and Caesalpiniaceae (*Cercidium praecox*) which has only one species of cactus, and a cactus-forest with 10 or more different species of cacti (Figs. 3.52B,C and 3.53, see also Figs. 3.6B). However, both types of forest mutually intermingle and form a mosaic-like pattern.

The climate at the sites of these thornbush-succulent forests is strongly seasonal and very dry during most of the year. The *Klimadiagramm* of Carora (Fig. 3.54) for example only shows a pronounced wet period from September to December and a very short rainy season in May to June. The open canopy of these deciduous forests allows much penetration of full sunlight, and the plants are subject to stress by

- high irradiance (hv),
- high temperature (T),
- scarcity of water or drought (H_2O)
- limited nutrient availability especially nitrogen (N).

Special ecophysiological adaptations have developed to these stresses.

Fig. 3.51. Thornbush-succulent forest in SW Madagascar with Didieraceae and Euphorbiaceae. (Photographs courtesy M. Kluge)

Fig. 3.52A-C

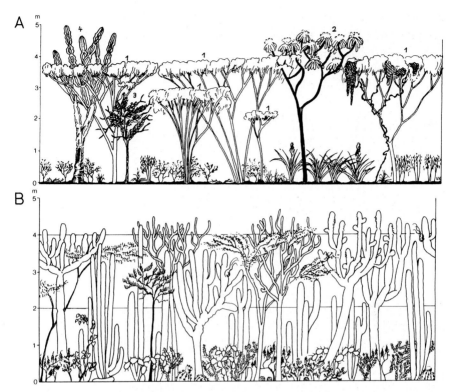

Fig. 3.53A, B. Transects of a thornbush forest (**A**) and a cactus forest (**B**) in the area of Carora, Venezuela. In **A** the trees are Mimosaceae (*1*), the Apocynaceae *Plumeria alba* (*2*), and the Cactaceae *Pereskia guamacho* (*3*). The only cactus is *Cereus jamacaru* (*4*). In **B** Cactaceae dominate the higher and lower strata. (Vareschi 1980, with kind permission of R. Ulmer)

Fig. 3.54.
Klimadiagramm of the thornbush-succulent forest site at Carora. (Vareschi 1980, with kind permission of R. Ulmer)

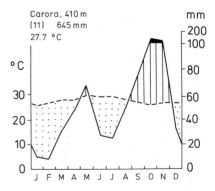

◀**Fig. 3.52A-C.** Caatinga in the state of Ceará, Brazil (**A**), and thornbush-succulent forests near Carora, Venezuela (**B, C**)

3.7.2
Interaction of Stress Factors and the Midday-Depression in Plants with C_3-Photosynthesis

The stressors mentioned above (Sect. **3.7.1**) may interact as follows:

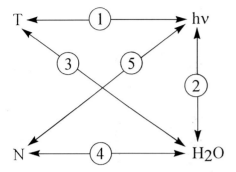

1. hv $\leftrightarrow$ T: Absorption of radiation by leaves leads to heating.
2. hv $\leftrightarrow$ H_2O: Heating and drying of the atmosphere, increases the leaf-air water-vapour pressure gradient and thus leads to increased transpirational water loss.
3. T $\leftrightarrow$ H_2O: Water loss can be controlled by closure of stomata, but this then reduces transpirational cooling by evaporation, and leaves heat up further.
4. H_2O $\leftrightarrow$ N: Soil water deficit reduces the availability of N; the transpiration stream serves distribution of nutrients in the plant.
5. N $\leftrightarrow$ hv: Mineral nutrition, particularly nitrogen supply, is important for constituting and repairing the photosynthetic apparatus (Sect. **3.6.2.2**).

In their reactions to these interacting stress effects, the plants must optimise responses to any one particular limitation, which may lead to disadvantages; for example closing stomata at high irradiance and low water availability reduces water loss but also causes more heating; CO_2-uptake is prevented and this leads to the dangers of photoinhibition.

The performance of a plant of the C_3-bromeliad *Pitcairnia integrifolia*, which grows in the thornbush-forest of Trinidad and smaller adjacent islands, demonstrates the effects of these interactions when studied on a clear and hot day (Lüttge et al. 1986b; Fig. **3.55**).

Photosynthetic CO_2-uptake rose after dawn as light-intensity increased and reached the highest rate at about 09.00 h. During this time temperature increased from about 23 °C to about 36 °C, but leaf temperature remained very close to air temperature. Beyond that point stomata began to shut and had fully closed by noon, when leaf conductivity to water vapour, g_{H2O}, was zero (not shown in Fig. **3.55**). At this time, and until about 15.00 h irradiance had attained its highest level around 2000 µmol photons m^{-2} s^{-1} and leaf temperature now increased much above air temperature with the highest value close to 52 °C and almost 8 °C higher than air temperature. If inhibition of CO_2-uptake was only due to stomatal closure, one would have expected intercellular CO_2-concentration (p^i_{CO2}) to have remained at low levels during this period. However, p^i_{CO2} rose and this shows that there were likely to be photoinhibitory responses occurring as well as the well docu-

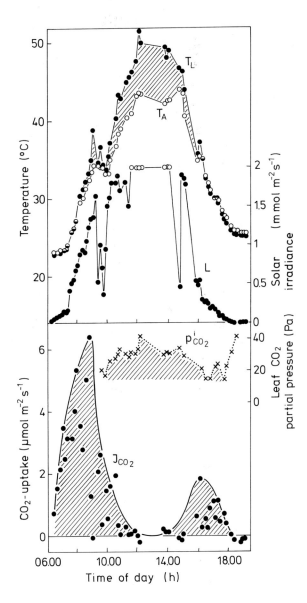

Fig. 3.55.
Daily course of CO_2 uptake (J_{CO_2}), intercellular CO_2 concentration ($p^i_{CO_2}$), leaf temperature (T_L), air temperature (T_A) and solar irradiance (L) for a plant of *Pitcairnia integrifolia* in Trinidad and a plant with the head of the porometer attached (photograph see page 124), which was used for measurements. (Lüttge et al. 1986 b)

Fig. 3.55 (continued)

mented change in carboxylation efficiency at this time. Later in the afternoon, when irradiance and temperatures declined again, stomata re-opened and p^i_{CO2} dropped, but CO_2-uptake only reached less than a third of the rate attained in the morning. Hence, strain during the hottest time of the day was only partially elastic and had a strong plastic component. Only during the subsequent night water uptake and rehydration as well as possible repair mechanisms may restore photosynthetic capacity.

The phenomenon of reduced gas exchange during the hottest time of the day is called **midday-depression**. It is very frequent among trees, shrubs and herbs in hot and arid regions (Schulze et al. 1974, 1975a,b; Tenhunen et al. 1980, 1981, 1984). The midday-depression may be smaller or larger. Gas exchange may be totally absent during this time by full stomatal closure. Moreover, recovery in the afternoon, as shown for example in Fig. **3.55**, may be expressed to different extents and with increasing drought it may not occur at all. Usually nocturnal rehydration may provide more effective recovery but this as well will be reduced as drought becomes increasingly severe.

In addition to mechanisms of stress tolerance, there are also means of stress avoidance. *Pitcairnia integrifolia* for example may roll its leaves, exposing only the lower abaxial surface to the sun. This surface is densely covered by silvery trichomes composed of dead cells which effectively reflect the light. Compared to white paper (100 % reflectance) the reflectance of the abaxial leaf surface with scales was found to be 46.5 % but only 19.8 % when the scales were removed.

3.7.3
Escape from the Dilemma Desiccation or Starvation: Crassulacean Acid Metabolism

Choosing between limiting the effects of any one stress represents a daily "damage limitation excercise" such that plants with C_3-photosynthesis face the dilemma of desiccation or starvation, when under water stressed conditions. With the midday-depression, the strategy is to try to avoid desiccation by stomatal closure at the expense of CO_2-supply for photosynthesis. Desiccation is always more rapid and is the more immediate danger than starvation. One escape from this dilemma is provided by the evolution of crassulacean acid metabolism (CAM) (Box **3.11**), where CO_2 is fixed during the night, when water-vapour pressure saturation deficit of the atmosphere is much lower than during the day, and hence stomatal opening has a smaller effect on the water budget of the plants. The CO_2 fixed is stored in chemical form as malic acid, remobilized again during the day and made available for photosynthesis, so that the plants can utilize the light energy of solar irradiance for CO_2-assimilation behind closed stomata.

This mode of photosynthesis was first discovered in plants of the genus *Kalanchoë* (see Sect. **2.5**), which belong to the family of the Crassulaceae, and hence the name. However, it must have evolved independently several times, i.e. polyphyletically, since there are CAM-performing taxa on almost all branches of the phylogenetic tree of cormophytes (Fig. **3.56**). Among the plants of the thornbush-succulent forests many are CAM-plants, i.e. all the cacti in the new world, the succulent Euphorbiaceae in the old world, the Didieraceae and many of the rosette plants in the Bromeliaceae, Agavaceae and Liliaceae, to name the major ones.

However, CAM may not only operate in the simple day-night fashion described above. In fact it provides an enormous range of plasticity in form and function, allowing responses to environmental conditions to be optimised (see also Sect. **2.5**). The best way of describing these options is by reference to the four phases of CAM according to the nomenclature introduced by Osmond (1978; see Box **3.11**). Phase I represents nocturnal stomatal opening with CO_2-uptake, fixation and storage as malic acid, whereas during phase III daytime stomatal closure with CO_2-remobilization and assimilation occurs. Phases II and IV are transitional phases in the early morning and in the afternoon. Phase IV often plays an important role, because when CAM plants are well watered it may be quite extensive. Then CAM plants take up CO_2 directly from the atmosphere and assimilate it directly by the C_3-mode of photosynthesis via RUBISCO. This can make a major contribution to their productivity.

Conversely, water stress may become so severe that even CAM plants face the dilemma of desiccation or starvation. Then, stomata may be closed even during the night, and CAM represents an option for survival by recycling CO_2 internally. The CO_2 evolved nocturnally during respiratory metabolism

Box 3.11

Crassulacean acid metabolism (CAM)

In CAM plants there are two ways of primary CO_2 fixation, namely via the enzymes phosphoenolpruvate-carboxylase (PEPC) and ribulosebisphosphate-carboxylase oxygenase (RUBISCO). In its typical performance CAM has four phases (Osmond 1978):

- Phase I:
  Nocturnal dark fixation of CO_2 via PEPC generating malic acid, which is translocated into the vacuole by proton pumps (H^+-ATPase and H^+-pyrophosphatase - PP_iase - transporting protons) and a malate transporter (transporting malate^{2-}) at the tonoplast.

- Phase II:
  A transition phase in the early morning, after light energy becomes available, with primary CO_2 fixation partially via PEPC and RUBISCO respectively.

- Phase III:
  Efflux of non-dissociated malic acid from the vacuole, malate decarboxylation and refixation of the CO_2 via RUBISCO behind closed stomata.

- Phase IV:
  Opening of stomata in the afternoon, when nocturnally accumulated malic acid is consumed, and primary CO_2 fixation via RUBISCO.

CAM may play a role as a water-conserving mechanism at different levels of drought stress.

- In the typical performance dominating nocturnal CO_2 uptake reduces transpirational loss of water related to CO_2 acquired and thus increases water-use efficiency, because the evaporative demand on leaves with open stomata is smaller in the dark than in the light.

- At increased drought stress first phase IV and then also phase II are eliminated, and stomata remain closed for the whole light period, further restricting transpirational loss of water.

- At still more severe drought, stomata may also be partially or totally closed during the dark period. In this situation the CO_2 fixed nocturnally for the accumulation of malic acid partially or totally may come from internal sources, i.e. mainly respiration (CO_2 recycling). This further reduces transpirational loss of water but also limits carbon acquisition.

Box 3.11 (Continued)

The scheme of CAM shows the key reactions in metabolism. With *PYR* pyruvate; *PEP* phosphoenolpyruvate; *OAA* oxaloacetate; *MAL* malate; P_i inorganic phosphate; *[CH₂O]* carbohydrate and transport across the tonoplast: with *MC* malate transporter; the H⁺-ATPase and H⁺-PP$_i$ase, and passive malic acid efflux.

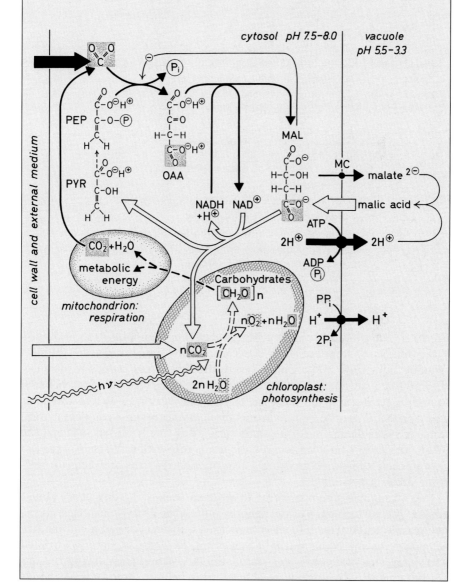

Box 3.11 (Continued)

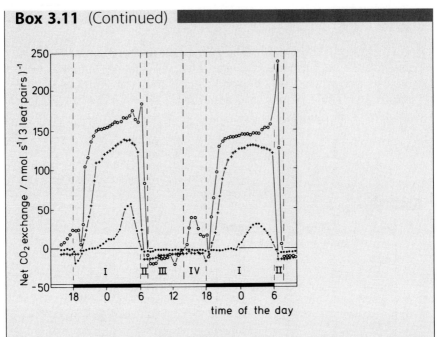

Net CO_2 exchange by the CAM-plant *Kalanchoë daigremontiana* with increasing drought stress: o—o well-watered; +—+ low and ●—● high drought stress. Phases I to IV are indicated. Phase II and IV CO_2 exchange is expressed only in the well-watered plant; onset of phase I CO_2 exchange is delayed in the severely stressed plant. (Smith and Lüttge 1985).

is refixed and stored as malic acid; the day-time remobilization and reassimilation, using solar radiation, recycles carbohydrate reserves for the subsequent night (Box **3.11**). Under severe drought stress cacti, for instance, keep stomata closed continuously for many months (see also Sect. **6.3.2.1**). By CO_2-recycling they do not gain carbon, but very little is lost and solar energy can be used to maintain metabolism and remain competent until water is available again. At the same time, with totally closed stomata, the plants lose only a little water via cuticular transpiration. Water storage tissues in cacti and other succulents also provide reserves and help to overcome drought periods.

In a drought deciduous forest in western Mexico Lerdau et al. (1992) studied the performance of the arborescent cactus *Opuntia excelsea*. In the dry season, when trees had shed their leaves, the cactus had a competitive advantage, as there was no light limitation. However, a factor associated with plant size, possibly water status, limited carbon gain during the dry season. Larger individuals were able to utilize water stored in their trunks

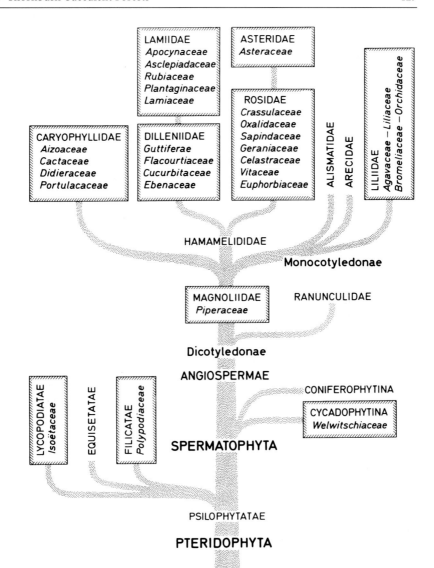

Fig. 3.56. Phylogenetic tree of taxa with Crassulacean acid metabolism indicated *inside boxes*. (Lüttge 1987)

and main branches (see also Sect. **6.3.2.1**). Light availability in the forest understory constrained CO_2-assimilation of the cactus in the wet season.

Daytime CO_2-remobilization from nocturnally stored organic acids also reduces the danger of photoinhibition, which otherwise would occur due to low internal CO_2-levels at high irradiance (Osmond 1982; Adams and Osmond 1988; Griffiths 1989). In fact internal CO_2-levels (p^i_{CO2}) behind

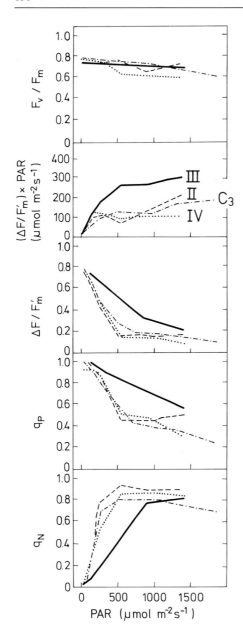

Fig. 3.57.
Chlorophyll-fluorescence variables
(see Box **3.8** for explanation) in
Clusia minor in the C$_3$ state (–·–·–)
and in the CAM state (– – – phase II,
····· phase IV, —— phase III).
(Haag-Kerwer and Lüttge 1995)

closed stomata during the light period of CAM may be very high and reach
up to a few percent (Cockburn et al. 1979; Kluge et al. 1981; see Lüttge
1987). Experiments measuring chlorophyll fluorescence with the facultative
CAM-tree *Clusia minor*, which can perform both CAM and C$_3$-
photosynthesis (Sect. **4.4.1**), have shown that photoinhibition, if it occurs, is

likely to be observed during phase IV of CAM, when stomata are open and plants fix CO_2 via RUBISCO rather than in phase III. Light response characteristics of chlorophyll-fluorescence variables (see Box **3.8**) in phases II and IV were similar to those observed with *C. minor* in the C_3-state and very different to those of phase III of CAM (Haag-Kerwer and Lüttge 1995; Fig. **3.57**).

Of course, if nocturnal accumulation of malic acid occurs from recycled CO_2 alone (= 100 % recycling), this is an extreme case. However, stomata may only be partially closed during the night and malic acid accumulation may be due to both recycled CO_2 and CO_2-uptake from the atmosphere. Since the stoichiometry of CO_2-fixed to malic acid formed is unity, recycling can be calculated in absolute terms as

malic acid accumulated minus CO_2 taken up

or in relative terms (% recycling) as

$$\frac{\text{malic acid accumulated minus } CO_2 \text{ taken up}}{\text{malic acid accumulated}} \times 100.$$

The degree of recycling may then depend on the severity of drought stress. This is illustrated in Fig. **3.58** by a study of *Aechmea aquilega* and its higher altitude counterpart *Aechmea fendleri* along a gradient of altitude and precipitation in Trinidad. *A. aquilega* grows both terrestrially and epiphytically from very dry deciduous thornbush-forests to quite wet forests, and *A. fendleri* is epiphytic in wet forests. Fig. **3.58** shows that with increasing altitude and precipitation total CO_2-uptake by the *Aechmea*s increased and relative CO_2-recycling decreased considerably.

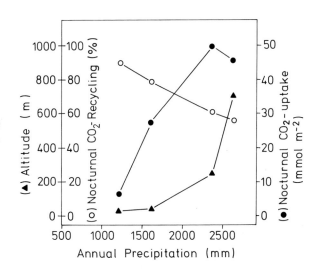

Fig. 3.58.
Net nocturnal CO_2 uptake from the atmosphere and internal CO_2 recycling of *Aechmea* (*A. aquilega* at the three lower altitudes and *A. fendleri* at the highest altitude) in relation to altitude and precipitation in Trinidad. (After data of Griffiths et al. 1986)

3.8
Summary: Levels of Complexity and Diversity in Tropical Forests

In this Chap. we have been enlightened by the richness of tropical forests, where we encounter a high degree of diversity on all levels of complexity and integration, i.e.

- on the synecological level of ecosystems with the different types of forests (Sect. **3.2**);
- on the floristic level (Sect. **3.3.1**);
- on the level of plant types and life forms (Sect. **3.4**);
- on the ecophysiological or autecological level (Sects. **3.6** and **3.7**).

Climatic, geomorphological and geographical conditions determine the diversity of forest types. Interactions of individual environmental factors and a certain degree of stress lead to expression of phenotypic plasticity, which is one of the bases for development of life-forms and evolution of species. Physiological, biochemical and molecular mechanisms provide the functional basis.

References

Adams WW, Osmond CB (1988) Internal CO_2 supply during photosynthesis of sun and shade grown CAM plants in relation to photoinhibition. Plant Physiol 86:117–123

Anderson JM, Thomson WW (1989) Dynamic molecular organization of the plant thylakoid membrane. Photosynthesis. Alan R Liss, New York, pp 161–182

Beard JS (1946) The natural vegetation of Trinidad. Oxford Forestry Memoirs, No 20. Oxford University Press, Oxford

Beard JS (1955) The classification of tropical American vegetation types. Ecology 36:89–100

Beck E, Lüttge U (1990) Streß bei Pflanzen. Biol Unserer Zeit 20:237–244

Bell G, Lechowicz MJ, Appenzeller A, Chandler M, DeBlois E, Jackson L, Mackenzie B, Preziosi R, Schallenberg M, Tinker N (1993) The spatial structure of the physical environment. Oecologia 96:114–121

Bilger W, Björkman O (1994) Relationships among violaxanthin deepoxidation, thylakoid membrane conformation, and non-photochemical chlorophyll fluorescence quenching in leaves of cotton (*Gossypium hirsutum* L.). Planta 193:238–246

Büch K, Stransky H, Bigus H-J, Hager A (1994) Enhancement by artificial electron acceptors of thylakoid lumen acidification and zeaxanthin formation. J Plant Physiol 144:641–648

Chazdon RL, Fetcher N (1984) Light environments of tropical rainforests. In: Medina E, Mooney HA, Vázquez-Yanes C (eds) Physiological ecology of plants in the wet tropics. Dr W Junk, The Hague, pp 27–50

Choong MF, Lucas PW, Ong JSY, Pereira B, Tan HTW, Turner IM (1992) Leaf fracture toughness and sclerophylly:their correlations and ecological implications. New Phytol 121:597–610

Chow WS, Qian L, Goodchild DJ, Anderson JM (1988) Photosynthetic acclimation of *Alocasia macrorrhiza* (L.) G. Don. to growth irradiance: structure, function and composition of chloroplasts. Aust J Plant Physiol 15:107–122

Clements FE (1936) Nature and structure of the climax. J Ecol 24:252–284

Cockburn W, Ting IP, Sternberg LO (1979) Relationships between stomatal behaviour and the internal carbon dioxide concentrations in crassulacean acid metabolism plants. Plant Physiol 63:1029–1032

Cramer F (1993) Chaos and order. The complex structure of living systems. VCH, Weinheim

Critchley C, Russell AW (1994) Photoinhibition of photosynthesis in vivo:the role of protein turnover in photosystem II. Physiol Plant 92:188–196

Davidson DW, Epstein WW (1989) Epiphytic associations with ants. In: Lüttge U (ed) Vascular plants as epiphytes. Evolution and ecophysiology. Ecological Studies, vol 76. Springer, Berlin Heidelberg New York, pp 200–233

Demmig-Adams B (1990) Carotenoids and photoprotection:a role for the xanthophyll zeaxanthin cycle. Biochim Biophys Acta 1020:1–24

Demmig-Adams B, Adams WW (1992) Photoprotection and other responses of plants to high light stress. Annu Rev Plant Physiol Plant Mol Biol 43:599–626

Demmig-Adams B, Adams WW (1993) The xanthophyll cycle, protein turnover and the high light tolerance of sun-acclimated leaves. Plant Physiol 103:1413–1420

Demmig-Adams B, Adams WW (1994) Capacity for energy dissipation in the pigment bed in leaves with different xanthophyll cycle pools. Aust J Plant Physiol 21:575–588

Doley D, Yates DJ, Unwin GL (1987) Photosynthesis in an Australian rainforest tree, *Argyrodendron peralatum*, during the rapid development and relief of water deficits in the dry season. Oecologia 74:441–449

Doley D, Unwin GL, Yates DJ (1988) Spatial and temporal distribution of photosynthesis and transpiration by single leaves in a rainforest tree, *Argyrodendron peralatum*. Aust J Plant Physiol 15:317–326

Evans JR (1988) Acclimation by the thylakoid membranes to growth irradiance and the partitioning of nitrogen between soluble and thylakoid proteins. Aust J Plant Physiol 15:93–106

Fetene M, Lee HSJ, Lüttge U (1990) Photosynthetic acclimation in a terrestrial CAM bromeliad, *Bromelia humilis* Jacq. New Phytol 114:399–406

Field CB (1988) On the role of photosynthetic responses in constraining the habitat distribution of rainforest plants. Aust J Plant Physiol 15:343–358

Flores S (1992) Growth and seasonality of seedlings and juveniles of primary species of a cloud forest in northern Venezuela. J Trop Ecol 8:299–305

Griffiths H (1989) Carbon dioxide concentrating mechanisms and the evolution of CAM in vascular epiphytes. In: Lüttge U (ed) Vascular plants as epiphytes:evolution and ecophysiology. Ecological studies, vol 76. Springer, Berlin Heidelberg New York, pp 42–86

Griffiths H, Lüttge U, Stimmel K-H, Crook CE, Griffiths NM, Smith JAC (1986) Comparative ecophysiology of CAM and C_3 bromeliads. III. Environmental influences on CO_2 assimilation and transpiration. Plant Cell Environ 9:385–393

Grime JP, Mackey JML, Hillier SH, Read DJ (1987) Floristic diversity in a model system using experimental microsoms. Nature 328:420–422

Haag-Kerwer A, Lüttge U (1995) Differential responses of chlorophyll fluorescence to light intensity during the distinguished phases of Crassulacean acid metabolism in comparison to C_3 photosynthesis in the C_3/CAM-intermediate plant *Clusia minor* L. (in preparation; see Haag-Kerwer A (1994) Photosynthetische Plastizität bei *Clusia* und *Oedematopus*. Dr. rer.-nat.-Thesis, Darmstadt)

Hager A (1980) The reversible, light-induced conversions of xanthophylls in the chloroplast. In: Czygan FC (ed) Pigments in plants. G Fischer, Stuttgart, pp 57–79

Hager A, Holocher K (1994) Localization of the xanthophyll-cycle enzyme violaxanthin de-epoxidase within the thylakoid lumen and abolition of its mobility by a (light-dependent) pH decrease. Planta 192:581–589

Hastings A, Hom CL, Ellner S, Turchin P, Godfray HCJ (1993) Chaos in ecology:is mother nature a strange attractor? Annu Rev Ecol Syst 24:1–33

Hoffmeister J (1955) Wörterbuch der philosophischen Begriffe. Hamburg (cited after Vareschi 1980)

Horton P, Ruban AV, Walters RG (1994) Regulation of light harvesting in green plants. Indication by nonphotochemical quenching of chlorophyll fluorescence. Plant Physiol 106:415–420

Huber O (1978) Light compensation point of vascular plants of a tropical cloud forest and an ecological interpretation. Photosynthetica 12:382–390

Hübinger B, Doerner R, Martienssen W (1993) Local control of chaotic motion. Z Physik B 90:103–106

Huc R, Ferhi A, Guehl JM (1994) Pioneer and late stage tropical rainforest tree species (French Guiana) growing under common conditions differ in leaf gas exchange regulation, carbon isotope discrimination and leaf water potential. Oecologia 99:297–305

Humbeck K, Senger H (1984) The blue light factor in sun and shade plant adaptation. In: Senger H (ed) Blue light effects in biological systems. Springer, Berlin Heidelberg New York, pp 344–351

Jacobs M (1988) The tropical rain forest. Springer, Berlin Heidelberg New York

Junk WJ, Furch K (1985) The physical and chemical properties of Amazonian waters and their relationships with biota. In: Prance GT, Lovejoy TE (eds) Amazonia. Pergamon, Oxford, p 7

Kirschbaum MUF, Pearcy RW (1988a) Concurrent measurements of oxygen- and carbon-dioxide exchange during light flecks in *Alocasia macrorrhiza* (L.) G. Don. Planta 174:527–533

Kirschbaum MUF, Pearcy RW (1988b) Gas exchange analysis of the fast phase of photosynthetic induction in *Alocasia macrorrhiza.* Plant Physiol 87:818–821

Kitajima K (1994) Relative importance of photosynthetic traits and allocation patterns as correlates of seedling shade tolerance of 13 tropical trees. Oecologia 98:419–428

Kluge M, Böhlke C, Queiroz O (1981) Crassulacean acid metabolism (CAM) in *Kalanchoë*. Changes in intracellular CO_2 concentration during continuous light or darkness. Planta 152:87–92

Kursar TA, Coley PD (1992) Delayed development of the photosynthetic apparatus in tropical rain forest species. Funct Ecol 6:411–422

Kursar TA, Coley PD (1993) Photosynthetic induction times in shade-tolerant species with long and short-lived leaves. Oecologia 93:165–170

Larcher W (1987) Streß bei Pflanzen. Naturwissenschaften 74:158–167

Lerdau MT, Holbrook NM, Mooney HA, Rich PM, Whitbeck JL (1992) Seasonal patterns of acid fluctuations and resource storage in the arborescent cactus *Opuntia excelsea* in relation to light availability and size. Oecologia 92:166–171

Levitt J (1980) Responses of plants to environmental stresses, vol I. Academic Press, New York

Lichtenthaler HK, Buschmann C, Döll M, Fietz H-J, Bach T, Kozel U, Meier D, Rahmsdorf U (1981) Photosynthetic activity, chloroplast ultrastructure and leaf characteristics of high-light and low-light plants and of sun and shade leaves. Photosynth Res 2:115–141

Linsenmair KE (1995) Biologische Vielfalt und ökologische Stabilität. In: Markl H, Geiler G, Großmann S, Oesterhelt D, Schmidbaur H, Quadbeck-Seeger HJ, Truscheit E (eds) Wissenschaft in der globalen Herausforderung. Verh Ges Dtsch Naturforsch Ärzte, 118 Vers Hamburg, Wiss Verlagsgesellschaft, Stuttgart, pp 267–295

Lloyd AL, Lloyd D (1995) Chaos. Its significance and detection in biology. Biological Rhythm Res 26:233–252

Lovelock CE, Jebb M, Osmond CB (1994) Photoinhibition and recovery in tropical plant species:response to disturbance. Oecologia 97:297–307

Lüttge U (1985) Epiphyten:Evolution und Ökophysiologie. Naturwissenschaften 72:557–566

Lüttge U (1987) Carbon dioxide and water demand:crassulacean acid metabolism (CAM) a versatile ecological adaptation exemplifying the need for integration in ecophysiological work. New Phytol 106:593–629

Lüttge U (1995) Ecophysiological basis of the diversity of tropical plants: The example of the genus *Clusia*. In: Heinen HD, San José JJ, Caballero-Arias H (eds) Nature and human ecology in the neotropics. Sci Guaianae 5:23–26

Lüttge U, Ball E, Kluge M, Ong BL (1986a) Photosynthetic light requirements of various tropical vascular epiphytes. Physiol Vég 24:315–331

Lüttge U, Klauke B, Griffiths H, Smith JAC, Stimmel K-H (1986b) Comparative ecophysiology of CAM and C$_3$ bromeliads. V. Gas exchange and leaf structure of the C$_3$ bromeliad *Pitcairnia integrifolia*. Plant Cell Environ 9:411–419

Lüttge U, Kluge M, Bauer G (1994) Botanik, 2nd edn. VCH, Weinheim

Mattheck C (1992) Design in der Natur. Der Baum als Lehrmeister. Rombach Freiburg i. Breisgau

Maxwell C, Griffiths H, Young AJ (1994) Photosynthetic acclimation to light regime and water stress by the C$_3$-CAM epiphyte *Guzmania monostachia*:gas exchange characteristics, photochmical efficiency and the xanthophyll cycle. Funct Ecol 8:746–754

Maxwell C, Griffiths H, Borland AM, Young AJ, Broadmeadow MSJ, Fordham MC (1995) Short-term photosynthetic responses of the C$_3$-CAM epiphyte *Guzmania monostachia* var. *monostachia* to tropical seasonal transitions under field conditions. Aust J Plant Physiol 22:771–781

May RM (1976) Simple mathematical models with very complicated dynamics. Nature 261:459–467

Medina E (1983) Adaptations of tropical trees to moisture stress. In: Golley FB (ed) Tropical rain forest ecosystems, A. Structure and function. Elsevier Amsterdam, pp 225–237

Medina E (1986) Forests, savannas and montane tropical environment. In: Baker NR, Long SP (eds) Photosynthesis in contrasting environments. Elsevier Amsterdam, pp 139–171

Medina E, Cuevas E (1989) Patterns of nutrient accumulation and release in Amazonian forests of the upper Rio Negro basin. In: Proctor J (ed) Mineral nutrients in tropical forest and savanna ecosystems. Blackwell Oxford, pp 217–240

Medina E, Cuevas E (1994) Mineral nutrition:humid tropical forests. Prog Bot 55:115–129

Medina E, Klinge H (1983) Productivity of tropical forests and tropical woodlands. In: Lange OL, Nobel PS, Osmond CB, Ziegler H (eds) Physiological plant ecology. IV. Ecosystem Processes:Mineral Cycling, Productivity and Man's Influence. Springer, Berlin Heidelberg New York, pp 281–303

Medina E, Montez G, Cuevas E, Roksandic Z (1986) Profiles of CO$_2$ concentration and δ^{13}C values in tropical rain forests of the upper Rio Negro basin, Venezuela. J Trop Ecol 2:207–217

Morawetz W, Wallnöfer B (1992) Die Ameisenpflanzen entlang eines Transekts durch das Sira-Gebirge (Peruanisches Amazonien) und ihre ökologische Stellung im Regenwald. Dtsch Ges Tropenökologie, Jahrestagung Bonn 1992

Morawetz W, Henzl M, Wallnöfer B (1992) Tree killing by herbicide producing ants for the establishment of pure *Tococa occidentalis* populations in the Peruvian Amazon. Biodivers Conserv 1, 19–33

Murali KS, Sukumar R (1993) Leaf flushing phenology and herbivory in a tropical dry deciduous forest, southern India. Oecologia 94:114–119

Newell EA, McDonald EP, Strain BR, Denslow JS (1993) Photosynthetic responses of *Miconia* species to canopy openings in a lowland tropical rainforest. Oecologia 94:49–56

Nickol MG (1992) Untersuchungen der Myrmekodomatien von *Tococa guianensis* (Melastomataceae). Dtsch Ges Tropenökologie, Jahrestagung Bonn 1992

Noss RF (1983) A regional landscape approach to maintain diversity. BioScience 33:700–706

Osmond CB (1978) Crassulacean acid metabolism:a curiosity in context. Annu Rev Plant Physiol 29:379–414

Osmond CB (1982) Carbon cycling and stability of the photosynthetic apparatus in CAM. In: Ting IP, Gibbs M (eds) Crassulacean acid metabolism. American Society of Plant Physiologists, Rockville, pp 112–127

Osmond CB, Grace SC (1995) Perspectives on photoinhibition and photorespiration in the field:quintessential inefficiencies of the light and dark reactions of photosynthesis? J Exp Bot 46:1351–1362

Pearcy RW, Osteryoung K, Calkin HW (1985) Photosynthetic responses to dynamic light environments by Hawaiian trees. Plant Physiol 79:896–902

Pfündel E, Bilger W (1994) Regulation and possible function of the violaxanthin cycle. Photosynth Res 42:89–109

Poorter L, Oberbauer SF (1993) Photosynthetic induction responses of two rainforest tree species in relation to light environment. Oecologia 96:193–199

Reich PB, Uhl C, Walters MB, Ellsworth DS (1991) Leaf life span as a determinant of leaf structure and function among 23 amazonian tree species. Oecologia 86:16–24

Reich PB, Walters MB, Ellsworth DS, Uhl C (1994) Photosynthesis-nitrogen relations in Amazonian tree species. I. Patterns among species and communities. Oecologia 97:62–72

Reichholf JH (1994) Biodiversity. Why are there so many different species? Universitas 1994/1:42–51

Remmert H (1985) Was geschieht im Klimax-Stadium? Ökologisches Gleichgewicht durch Mosaik aus desynchronen Zyklen. Naturwissenschaften 72:505–512

Remmert H (1991) The mosaic cycle of ecosystems. Ecological Studies, vol 85. Springer, Berlin Heidelberg New York

Riddoch I, Lehto T, Grace J (1991) Photosynthesis of tropical tree seedlings in relation to light and nutrient supply. New Phytol 119:137–147

Sarmiento G (1984) The ecology of neotropical savannas. Harvard University Press, Cambridge

Sassenrath-Cole GF, Pearcy RW (1992) The role of ribulose-1,5-bisphosphate regeneration in the induction requirement of photosynthetic CO_2 exchange under transient light conditions. Plant Physiol 99:227–234

Schäfer C, Schmid V (1993) Pflanzen im Lichtstreß. Biol Unserer Zeit 23:55–62

Schindler C, Lichtenthaler HK (1996) Photosynthetic CO_2-assimilation, chlorophyll fluorescence and zeaxanthin accumulation in field grown maple trees in the course of a sunny and a cloudy day. J Plant Physiol 148:399–412

Schreiber U, Bilger W (1987) Rapid assessment of stress effects on plant leaves by chlorophyll fluorescence measurements. In: Tenhunen JD, Catarino FM, Lange OL, Oechel WC (eds) Plant responses to stress. Functional analysis in Mediterranean ecosystems. NATO-ASI-Series G, Ecological sciences, vol 15. Springer, Berlin Heidelberg New York, pp. 27–53

Schreiber U, Bilger W (1993) Progress in chlorophyll fluorescence research:major developments during the past years in retrospect. Prog Bot 54:151–173

Schulze E-D, Lange OL, Evenari M, Kappen L, Buschbom U (1974) The role of air humidity and leaf temperature in controlling stomatal resistance of *Prunus armeniaca* L. under desert conditions. I. A simulation of the daily course of stomatal resistance. Oecologia 17:159–170

Schulze E-D, Lange OL, Evenari M, Kappen L, Buschbom U (1975a) The role of air humidity and leaf temperature in controlling stomatal resistance of *Prunus armeniaca* L. under desert conditions. III. The effect on water use efficiency. Oecologia 19:303–314

Schulze E-D, Lange OL, Kappen L, Evenari M, Buschbom U (1975b) The role of air humidity and leaf temperature in controlling stomatal resistance of *Prunus armeniaca*

L. under desert conditions. II. The significance of leaf water status and internal carbon dioxide concentration. Oecologia 18:219–233

Schuster HG (1995) Deterministic chaos. VCH, Weinheim

Selye H (1973) The evolution of the stress concept. Am Sci 61:693–699

Sims DA, Pearcy RW (1994) Scaling sun and shade photosynthetic acclimation of *Alocasia macrorrhiza* to whole-plant performance. – I. Carbon balance and allocation at different daily photon flux densities. Plant Cell Environ 17:881–887

Sims DA, Gebauer RLE, Pearcy RW (1994) Scaling sun and shade photosynthetic acclimation of *Alocasia macrorrhiza* to whole-plant performance. – II. Simulation of carbon balance and growth at different photon flux densities. Plant Cell Environ 17:889–900

Smith H, Samson G, Fork DC (1993) Photosynthetic acclimation to shade:probing the role of phytochromes using photomorphogenic mutants of tomato. Plant Cell Environ 16:929–937

Smith JAC, Lüttge U (1985) Day-night changes in leaf water relations associated with the rhythm of crassulacean acid metabolism in *Kalanchoë daigremontiana*. Planta 163:272–282

Sobrado MA (1991) Cost-benefit relationships in deciduous and evergreen leaves of tropical dry forest species. Func Ecol 5:608–616

Sobrado MA (1995) Seasonal differences in nitrogen storage in deciduous and evergreen species of a tropical dry forest. Biol Plant 37:291–295

Stephens GR, Waggoner PE (1970) Carbon dioxide exchange of a tropical rainforest. Part I. BioScience 20:1050–1053

Stewart GR, Hegarty EE, Specht RL (1988) Inorganic nitrogen assimilation in plants of Australian rainforest communities. Physiol Plant 74:26–33

Stewart GR, Joly CA, Smirnoff N (1992) Partitioning of inorganic nitrogen assimilation between the roots and shoots of cerrado and forest trees of contrasting plant communities of South East Brasil. Oecologia 91:511–517

Stone L, Ezrati S (1996) Chaos, cycles and spatiotemporal dynamics in plant ecology. J Ecol 84:279–291

Strauss-Debenedetti S, Bazzaz FA (1991) Plasticity and acclimation to light in tropical Moraceae of different successional positions. Oecologia 87:377–387

Tenhunen JD, Lange OL, Braun M, Meyer A, Lösch R, Pereira JS (1980) Midday stomatal closure in *Arbutus unedo* leaves in a natural macchia under simulated habitat conditions in an environmental chamber. Oecologia 47:365–367

Tenhunen JD, Lange OL, Braun M (1981) Midday stomatal closure in mediterranean type sclerophylls under simulated habitat conditions in an environmental chamber. II. Effect of the complex of leaf temperature and air humidity on gas exchange of *Arbutus unedo* and *Quercus ilex*. Oecologia 50:5–11

Tenhunen JD, Lange OL, Gebel J, Beyschlag W, Weber JA (1984) Changes in photosynthetic capacity, carboxylation efficiency, and CO_2-compensation point associated with midday stomatal closure and midday depression of net CO_2 exchange of leaves of *Quercus suber*. Planta 162:193–203

Tilman D (1982) Resource competition and community structure. Princeton Univ. Press, Princeton

Tinoco-Ojanguren C, Pearcy RW (1992) Dynamic stomatal behaviour and its role in carbon gain during lightflecks of a gap phase and an understory *Piper* species acclimated to high and low light. Oecologia 92:222–228

Tinoco-Ojanguren C, Pearcy RW (1993a) Stomatal dynamics and its importance to carbon gain in two rainforest *Piper* species. I. VPD effects on the transient stomatal response to light flecks. Oecologia 94:388–394

Tinoco-Ojanguren C, Pearcy RW (1993b) Stomatal dynamics and its importance to carbon gain in two rainforest *Piper* species. II. Stomatal versus biochemical limitations during photosynthetic induction. Oecologia 94:395–402

Vareschi V (1980) Vegetationsökologie der Tropen. Ulmer, Stuttgart
Vázquez-Yanes C, Orozco-Segovia A (1993) Patterns of seed longevity and germination in
    the tropical rainforest. Annu Rev Ecol Syst 24:69–87
Walter H (1973) Vegetationszonen und Klima. Ulmer, Stuttgart
Walter H, Breckle S-W (1984) Ökologie der Erde, vol 2. Spezielle Ökologie der tropischen
    und subtropischen Zonen. G Fischer, Stuttgart
Watt AS (1947) Pattern and process in the plant community. J Ecol 35:1–22
West-Eberhard MJ (1986) Alternative adaptations, speciation, and phylogeny (a review).
    Proc Natl Acad Sci USA 83:1388–1392
West-Eberhard MJ (1989) Phenotypic plasticity and the origins of diversity. Annu Rev
    Ecol Syst 20:249–278
Whitmore TC (1990) An introduction to tropical rain forests. Oxford University Press,
    Oxford
Whittaker RH (1975) Communities and ecosystems, 2nd edn. Macmillan, New York

CHAPTER 4

Epiphytes, Lianas and Hemiepiphytes

4.1
The Conquest of Space

Perhaps epiphytism could be thought to be primarily the utilization of any possible surface for holdfast and establishment, i.e. a conquest of space with epiphytes found in aquatic and terrestrial habitats made up of various combinations of lower and higher plants. In aquatic habitats, i.e. lakes, rivers and the sea, there are always algae growing on each other. This not only applies to unicellular and filamentous forms and their colonies, but also to macroalgae like kelp and red algae. In the mesic terrestrial climate many **lower plants** are epiphytic, like mosses and lichens and also some forms of small pleurococcoid aerial green algae as well as cyanobacteria (blue green algae). In the tropics lower plants may constitute massive formations of epiphytic biomass, e.g. the mosses in upper montane cloud forests ("**moss-forests**", Fig. 4.1.).

Even the surfaces of leaves of plants in such forests may harbour a diverse flora or phyllosphere with bacteria, cyanobacteria, algae, mosses and lichens (Ruinen 1961, 1974; Coley et al. 1993) (Fig. 4.2). This can have adverse effects on light utilization by the host leaves. However, it has been suggested that nitrogen fixation by the many cyanobacteria among these **epiphylls** may also supplement nutrients available to the host (Ruinen 1965; Benzing 1990) or to the whole ecosystem. Dinitrogen (N_2) reduction in the phyllosphere mainly depends on light and water. In a premontane tropical rainforest in Costa Rica N-supply by N_2-fixation in the phyllosphere was calculated as $2-7$ kg N ha^{-1} yr^{-1} (Freiberg 1994) and the higher values ($30-60$ kg N ha^{-1} yr^{-1}) which have been presented in the literature for other areas are discussed critically (see also Sect. **7.3.1.3**)

If we exclude the mistletoes, which are true parasites, among the **vascular plants** in the temperate zone the fern *Polypodium vulgare* is the only known epiphyte, and moreover, it is only facultatively epiphytic. In contrast, the popular view of tropical rainforests is determined by the image of an abundant flora of epiphytes, vines and lianas, climbers with hanging and host-strangling shoots and curtains of aerial roots (Fig. **4.3**).

This image, although intuitively correct, needs to be carefully differentiated. Growth of lianas, climbers and vines is particularly rich at the perimeter of forests, along rivers, roads and around clearings. They are often light-demanding plants. Epiphytes are much more abundant in montane

Fig. 4.1. Upper montane cloud forest at 2000 m a.s.l. (Sierra Maigualida 05° 30'N, 65°15'W) with epiphytic mosses, "moss forest"

rainforests and in cooler upper montane fog and cloud forests, where air moisture is always high, than in the hot lowland rainforests.

Remembering that vascular plants evolved from aquatic ancestors during the conquest of land and that then they were subject to many new kinds of stress with respect to water and nutrient relations, it may not be surprising

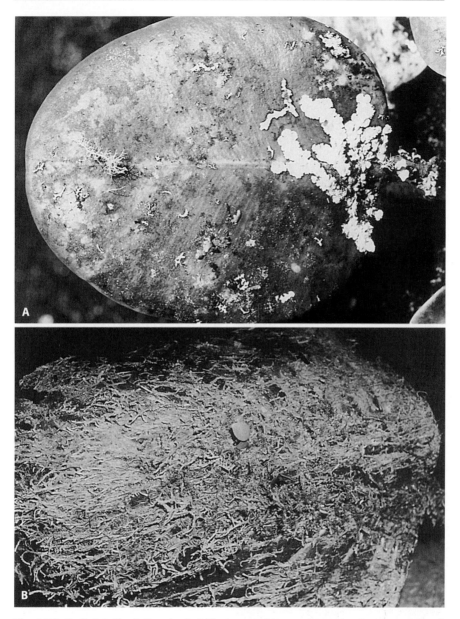

Fig. 4.2A, B. Epiphylls. **A** On a leaf of *Clusia* sp., with cyanobacteria, algae, mosses and lichens (Sierra Maigualida 05° 30'N, 65°15'W). **B** On a leaf in a cloud forest above Lake Coté (Costa Rica) with mosses and a higher-plant seedling

Fig. 4.3A-C. Rainforests with epiphytes and lianas. **A** Plate from Martius' Flora Brasiliensis (1840-1906). **B** A cloud forest above Lake Coté, Costa Rica. **C** Curtain of aerial roots of a strangler fig (*Ficus*) (Queensland, Australia)

that there is very little fossil record of epiphytism. Epiphytism must be a fairly recent development among vascular plants. Most epiphyte diversity dates from the Pliocene-Pleistocene (Benzing 1989a, 1990, Lüttge 1989). Against this background, when plants may even live in the air, this may be

Fig. 4.4 A, B. Atmospheric bromeliads of *Tillandsia flexuosa* (A) and *Tillandsia recurvata* (B) on telephone wires (Falcon State, Venezuela)

Table 4.1. Taxonomic diversity of epiphytes. (Data from Kress 1989)

	Number of taxa with vascular epiphytes	% epiphytes of total taxa of vascular plants
Species	23 466	10
Genera	879	7
Families	84	19
Orders	44	45

considered a rather extreme case of the conquest of space. The so-called **atmospheric bromeliads** constitute such life forms. They have given up any contacts with substrates supplying water and nutrients other than from the atmosphere. They may hang down from the branches of phorophytes (see Fig. **4.11 D** below) or may even get holdfast on wires of fences or telephone lines (Fig. **4.4**).

Evolution of epiphytism has clearly occurred many times and has been polyphyletic, since **taxonomic diversity** of epiphytes is quite substantial (Table **4.1**). The best-known families with epiphytes are headed by the Orchidaceae, although the Araceae, Piperaceae and Bromeliaceae are also important, and epiphytic taxa are abundant among ferns. In some tropical forests up to 50 % of all leaf biomass may be due to epiphytes, and of the known species of lianas 90 % are native to the tropics.

4.2
Lichens

As shown above (Sect. **4.1**, Fig. **4.1**) mosses and lichens may constitute a considerable floristic diversity and biomass among the epiphytes in tropical rainforests, especially in the cloud forests at higher elevations with their cooler nights (Seifriz 1924; Sipman 1989). Green and Lange (1994) recently have provided a comparison of photosynthesis in bryophytes and lichens. A major difference between the two groups is the effect of water relations on photosynthesis. In mosses, the CO_2-exchange surface is external, and the mosses have special water storage volumes, i.e. special cells – often dead cells – and capillary structures, which are separated from the gas exchange areas. Lichens have an internal CO_2-exchange surface with the phycobionts embedded in a relatively compact fungal tissue, and any water storage will tend to hinder gas exchange either within the compact tissue or at the outer lichen surface. Therefore, as compared to mosses, lichens tend to have lower maximal water content on a dry weight basis and there is the risk that high thallus water content impairs CO_2-uptake and assimilation. This difference between the two groups possibly explains the particular dominance of bryophytes in very wet habitats, while lichens can also be very successful in dry habitats.

Although floristics, taxonomy and habitat occupation by tropical lichens has been well studied (Galloway 1991), almost no work is available on their ecophysiology. Lichens in the temperate rainforests, e.g. in New Zealand, have been investigated to some extent (Green and Lange 1991; Green et al. 1991; Lange et al. 1993), but so far there are only a couple of ecophysiological studies on lichens in forests of the wet tropics (Lange et al. 1994; Zotz and winter 1994 a). Lange et al. 1994, have studied the gas exchange, water relations and potential productivity of the cyanobacterial basidiolichen *Dictyonema glabratum* living epiphytically, saxicolously and terrestrially in a lower montane tropical rainforest in Panama with an annual precipitation of 4000–4500 mm and a mean annual temperature of 21–22 °C (Fig. **4.5**). This lichen occurs both in shaded and exposed sites, and it is quite frequent in this forest. It has a rather unusual ecophyiological behaviour with respect to water saturation of its thallus and a number of additional traits, which explain its high productivity in the habitat.

Normally photosynthetic net CO_2-uptake is impaired in lichen thalli when they are oversaturated with water due to diffusion limitation from surface water or blocked air channels in the mycelium (e.g. see Sect. **8.2.2** and Fig. **8.16**). In contrast, *D. glabratum* maintains maximum rates of net-CO_2 uptake when it is fully hydrated up to a water content of 1000 % of its

Fig. 4.5. The lichen *Dictyonema glabratum* (syn. *Cora pavonia*) growing epiphytically on a 6-m-tall bush in a lower montane tropical rainforest in Panama. (Photograph courtesy B. Büdel; see Lange et al. 1994)

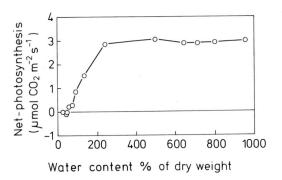

Fig. 4.6.
Net photosynthesis of the lichen *Dictyonema glabratum* in relation to water content in percent of dry weight. (Lange et al. 1994)

dry weight (Fig. **4.6**). This allows maximum benefit from the heavy rain storms occurring in the habitat. The lichen also possesses a mechanism for concentrating internal inorganic carbon by energy dependent transport, which occurs in many algae and also higher water plants (see Griffiths 1989; Badger et al. 1993). This mechanism allows photosynthesis at elevated intracellular CO_2-levels.

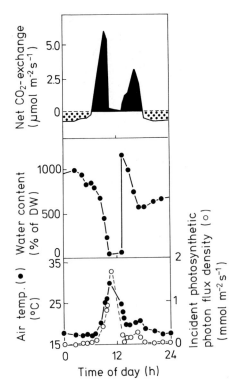

Fig. 4.7.
A daily course of net CO_2 exchange and water content of the lichen *Dictyonema glabratum* with air temperature and light intensity (PPFD) in its natural habitat in a lower montane rainforest in Panama; measured on 23 September 1993 by Lange et al. (1994)

Net CO_2-exchange on many days is typically bimodal with a peak in the morning, when thalli are wetted from dew and early fog, and another peak after midday when heavy showers may occur (Fig. **4.7**). In between, the thallus may dry out and CO_2-uptake ceases. In fact, drying out occurs for a few hours almost every day, and similar to most lichens, *D. glabratum* is also desiccation tolerant (Sect. **8.2.2**). However, unlike many other lichens it does not reactivate photosynthetic CO_2-fixation immediately after rewetting following desiccation, but it needs a recovery period of about 60 min. Thus it appears that lichens of the very moist lower montane rainforest show subtle changes in rehydration and reactivation charateristics as compared to lichens from temperate or arid habitats (Sect. **8.2.2**).

Another important trait is the thermophily of *D. glabratum*. It has been noted above that epiphytic mosses and lichens are particularly abundant in forests at higher elevations. The better supply of water in the epiphytic habitat by dew and fog may only be one reason for this distribution. An even more critical factor may be the reduction of respiration at lower night temperatures. Indeed, the cooler nights much reduce respiration and thus nocturnal loss of carbon, which is a considerable factor in overall productivity in the tropics and may be decisive for the general preponderance of epiphytic mosses and lichens in cloud and fog forests. In *D. glabratum* respiration increases only slightly with temperature up to 22 °C, but then increases sharply with higher temperatures. Net photosynthesis increases up to 22 °C and then declines in parallel with increasing respiration, so that gross photosynthesis calculated from net photosynthesis and respiration remains at a constant high level up to 40 °C. The balance of net photosynthesis is positive

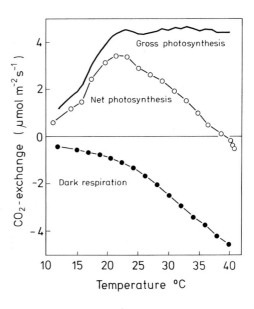

Fig. 4.8.
Net photosynthesis, dark respiration and calculated gross photosynthesis of the lichen *Dictyonema glabratum* in relation to temperature at a light intensity (PPFD) of 150 μmol m^{-2} s^{-1} and at high thallus water content. (Lange et al. 1994)

up to 35 °C (Fig. **4.8**). In fact the maximum rates of net CO_2-uptake in *D. glabratum* are quite high even in comparison to sun plants among vascular epiphytes (see below: Table **4.7**). On a thallus area basis the highest rate observed in the field was 8 μmol m^{-2} s^{-1}. Calculations have suggested that the annual relative production under the habitat conditions of *D. glabratum* in Panama is 2.28, i.e. a gain of 2.28 g of carbon per 1 g of initial thallus carbon. Thus, even with leaching of carbon under the influence of regular heavy rain which is frequently observed in lichens (Bruns-Strenge and Lange 1992), *D. glabratum* must retain sufficient surplus to allow rapid growth.

This behaviour of a tropical rainforest lichen is quite remarkable in comparison to the slow growth otherwise observed among lichens. The observations of Lange et al. (1994) put the biomass production of lichens and its ecological importance in tropical fog and cloud forests in a new perspective.

4.3
Life Forms of Vascular Plants in the Conquest of Phorophytes

4.3.1
Epiphytes

As noted above, the only vascular epiphyte in the mesic climate, the fern *Polypodium vulgare* is a **facultative epiphyte**. This means that there are gradations between the terrestrial and epiphytic habit (Gessner 1956; Richards 1952). In the tropics many species, e.g. among bromeliads and aroids, grow equally well terrestrially and epiphytically (Fig. **4.9**; see also below Table **4.4**).

Benzing (1989a) has conceived five different schemes alternatively classifying epiphytes in categories based on

I.   relationships to the host (or "phorophyte");
II.   growth habit;
III. humidity;
IV.   light;
V.   phorophyte-provided media

(Table **4.2**). The epiphytic **life form** is effectively encompassed in categories I and II, although the first goes much beyond epiphytes in a strict sense. The other three categories refer to the three major **stress factors** of epiphytic plant life (see Sect. **4.4.**). The entire system of five schemes is very useful as it offers a good summary of the great morphological and ecophysiological diversity among epiphytes and their associates. On the other hand, it suffers from the general problem of attempts of this kind of casting the

Fig. 4.9A, B. The bromeliad *Aechmea lingulata* growing both terrestrially (**A**) and epiphytically (**B**). (St. John-Island, US Virgin Islands, Lesser Antilles)

diversity of life into schematic systems (see also quotation of Hoffmeister (in Sect. **3.1**). Thus, the study of case stories may prove more appealing.

One of the most exciting case stories is offered by the **Bromeliaceae** (for a monograph see Martin 1994). They operate with **tanks** and **epidermal scales or trichomes**. The tanks are made up by densely overlapping leaf-bases of the rosette-forming bromeliads and depending on life form there is a gradation in effectiveness of water storing capacity. The scales are epidermal structures which developed increasing complexity during the evolution of bromeliads. They consist of living basal or foot cells, stalk cells, which may be living or dead in the mature stage of the trichomes, and the actual scales comprised of dead cells (Fig. **4.10**).

By the structure of tanks and scales we may distinguish 4 different **life forms of bromeliads** (Table **4.3**):

- Type I: **Soil Root.**
  Some bromeliads which are obligately terrestrial do not form tanks; often these forms are highly xeromorphic; they may be densely covered with scales; however the scales do not function in water and nutrient absorption but may rather serve reflection of light (see Sect. **3.7.2**; Fig. **3.55** and Fig. **4.11 A**).

Table 4.2. Five different schemes alternatively classifying epiphytes. (After Benzing 1989 a,b)

I.   **Relationships to the phorophyte**
    1. Autotrophs, using the phorophyte only for support
        1.1 Accidental
        1.2 Facultative
        1.3 Hemiepiphytic
            1.3.1 Primary
                  1.3.1.1 Strangling
                  1.3.1.2 Non-strangling
            1.3.2 Secondary
        1.4 Genuinely epiphytic
    2. Parasites

II. **Growth habit**
    1. Trees
    2. Shrubs
    3. Suffrutescent to herbaceous forms
        3.1 Tuberous
            3.1.1 Storage, woody and herbaceous
            3.1.2 Myrmecophytic, mostly herbaceous
        3.2 Broadly creeping: woody or herbaceous
        3.3 Narrowly creeping: mostly herbaceous
        3.4 Rosulate, herbaceous
        3.5 Root/leaf tangle, herbaceous
        3.6 Trash-basket, herbaceous

III **Humidity**
    1. Poikilohydrous (mostly lower plants)
    2. Homoiohydrous
        2.1 Hygrophytes
        2.2 Mesophytes
        2.3 Xerophytes
            2.3.1 Drought-endurers
            2.3.2 Drought-avoiders
        2.4 Impounders

IV. **Light**
    1. Exposure types
    2. Sun types
    3. Shade-tolerant types

V.   **Phorophyte-provided media**
    1. Relatively independent of rooting medium
        1.1 Atmospheric forms
        1.2 Twig and bark inhabitants
        1.3 Forms creating substitute soils or attracting ant colonies
    2. Utilizing preexisting specific rooting media
        2.1 Humus-dependent
            2.2.1 Shallow humus forms
            2.2.2 Deep humus forms
            2.2.3 Ant-nest garden and plant catchment inhabitants
        2.2 Parasites

Fig. 4.10A, B.
Schemes of scales of bro-
meliads. **A** Top view.
B Cross-section the living
cells of the scale *dotted*
and with a *nucleus*. The
black line along cells in **B**
indicates cutinization of
the epidermal cells and
the outer walls of the trichome stalk cells which
allows entry of solutes into
the leaves only via a specific pathway enforcing
membrane passage and
cytoplasmic control over
the solutes taken up.
(After Sitte 1991 with permission of G. Fischer-
Verlag)

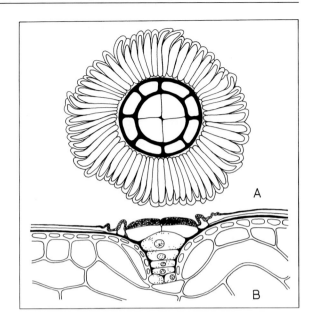

- Type II: **Tank Root.**
  Other obligately terrestrial bromeliads have rudimentary tanks, which
  have limited water and litter collecting capacity; the scales make only
  minor contributions to water and solute uptake; however, in addition to
  the soil-roots, plants of this type develop stem-borne "**tank-roots**"
  growing up between the overlapping leaf bases into the tanks (Fig. **3.36**
  and Fig. **4.11 B**).
- Type III: **Tank-Absorbing Trichome.**
  The roots have only mechanical functions in these epiphytic forms, the
  tanks effectively collect rain water and decomposing debris; scales are
  found most densely on the leaf bases in the tank, where they serve water
  and nutrient uptake (Fig. **3.51** and Figs. **4.9** and **4.11 C**).
- Type IV: **Atmospheric-Absorbing Trichome.**
  Tanks in these forms are mostly absent and only occasionally poorly
  developed; the entire leaf surface is covered by highly specialized scales,
  which provide the only route for uptake of water and minerals from rain
  and dust in the atmosphere; in some forms roots are lacking entirely
  (Figs.. **4.4** and **4.11 D**).

These life forms of bromeliads provide an interesting example of how the
vegetative plant form has been shaped by evolution towards epiphytism;
and in this case, particularly driven by the need for water and nutrient
acquisition in the epiphytic habitat.

Table 4.3. Life-forms of Bromeliaceae, their characterization and distribution among the tree subfamilies Pitcairnioideae, Bromelioideae and Tillandsioideae. (After Pittendrigh 1948; Smith et al. 1986a; Smith 1989)

Designation of life-form	Root system	Tank	Epidermal trichomes	Growth habit	Taxa
Type I	Soil roots	Lacking	Unspecialized and non-absorbent	Obligately terrestrial	Great majority of Pictarnioideae
Type II	Soil roots and tank roots	Rudimentary	Relatively unspecialized	Obligately terrestrial	All terrestrial Bromelioideae
Type III	Usually only mechanical	Well developed	Specialized and absorbent; concentrated on leaf base	Most obligately (some facultatively) epiphytic	All the epiphytic Bromelioideae, majority of Tillandsioideae
Type IV	Exclusively mechanical	Often entirely lacking	Specialized and absorbent; often cover entire shoot	Obligately epiphytic (or saxicolous)	Tillandsioideae: Several species of *Vriesea*, otherwise species exclusively *Tillandsia*

Fig. 4.11A-D. Life forms of bromeliads. **A** Type I, soil root, *Pitcairnia integrifolia*, Trinidad. **B** Type II, tank root, *Bromelia humilis*, Falcon, Venezuela. The basal leaves were removed and the rosette was turned upside down for photography, so that the tank roots growing upwards between the leaves can be seen. **C** Type III, tank-absorbing trichome, *Tillandsia fasciculata*, Cerro Santa Ana, Paraguana Peninsula, Falcon, Venzuela. **D** Type IV, atmospheric-absorbing trichome, *Tillandsia usneoides*, Merida, Venezuela

4.3.2
Lianas, Climbers, Vines and Hemiepiphytes

Lianas, climbers and vines are rooted in the soil and use other plants, especially trees, as support for growth away from the ground (Holbrook and Putz 1996a). It is mostly assumed that the particular advantage of this habit is to allow these plants to escape from deep shade and to reach the upper canopy of forests. This implies that their seeds would germinate in the shade and seedlings initially would grow upwards and develop in the shade. In the tropics, lush growth of lianas and climbers, however, is mostly found adjacent to clearings and in sites disturbed by man, and it appears that these forms need high irradiance for establishment and development. Perhaps, by clasping other plants, they are just saving investment in thick stems which would provide independent support for their biomass.

The plants climb using **tendrils** formed from modified leaves or parts of leaves, shoots or adventitious roots. Shoots wind around branches or form coils, which are then modified by the **secondary thickening,** so that the wood develops in the form of bands (Fig. **4.12**), or it is fragmented to individual strands forming rope- or cable-like structures which are resistant to torsion.

Fig. 4.12. Spirally twisted flat and band-like shoots of lianas

Box 4.1

Some basic principles of water relations of plants

The **water potential** ψ is defined as

- $$\psi = \frac{\mu_{H_2O} - \mu_{H_2O}^\circ}{\overline{V}_{H_2O}},$$

where $\mu_{H_2O}^\circ$ is the chemical potential of pure water, μ_{H_2O} the actual chemical potential of water in a solution and $\overline{V}_{H_2O}$ the partial molar volume of water.

ψ is best explained using an osmotic system of two chambers separated by a **semipermeable membrane** each having a vertical tube. The solvent particles ($\bullet$ = H_2O molecules) can pass the semipermeable membrane, whereas the solute particles (o = solute molecules) cannot permeate. The difference of **osmotic pressure** between the solutions of the two chambers, $\Delta\pi$, drives water across the semipermeable membrane from the chamber of the lower osmotic pressure, π, to the chamber of the higher π. The flow of water is a **volume flow J_v.** The associated volume change leads to the ascent of solution in the vertical tube of the chambers with the higher π. The column of solution exerts a **hydrostatic pressure** on the solution, ΔP, which counteracts the osmotic water flow driven by $\Delta\pi$. At thermodynamic equilibrium

- $\Delta P = \Delta\pi$ and $\Delta\psi = 0$,

i.e. there is no water potential difference between the two chambers since at any given time

- $\Delta\psi = \Delta P - \Delta\pi$.

All parameters have the physical unit of a pressure. π is related to the **solute concentration**, **c**, as follows

- $\pi = c \cdot R \cdot T$,

where T is the temperature in Kelvin and R the universal gas constant.

In plant cells P is built up at the elastic **cell walls** and is called **turgor pressure**.

Box 4.1 (Continued)

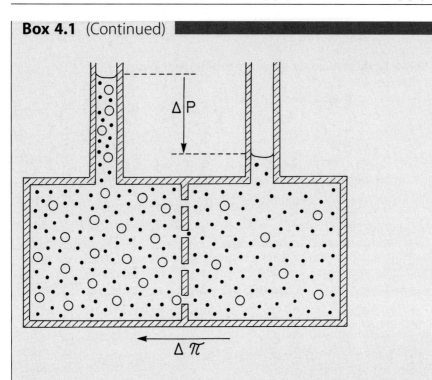

Measurements of ψ can be made by **psychrometric techniques** or using a **pressure chamber**. In the latter case a plant shoot or leaf is tightly sealed in the chamber, with the cut stem or petiole protruding to the outside. When pressure is exerted on the air in the chamber, xylem sap is expressed from the cut end. The equilibrium pressure, where the sap just reaches the cut end, can, with certain precautions, be related to the water potential of the stem or leaf.

P can be measured by inserting small glass capillaries into cells which are adjoined to a pressure transducing read-out system (**pressure probe**). Often it is also calculated from ψ and π, but this needs to be interpreted with care.

π is obtained from **psychrometry** or **freezing point determinations** of cell sap. **Plasmolysis** studies are also applied.

(See textbooks, e.g. Nobel 1983; Lüttge et al. 1994.)

Fig. 4.13A-D. Epiphytic seedlings of *Clusia rosea* in humus accumulation of tree forks (A) in an epiphytic garden (B), and inside tanks of the bromeliad *Aechmea lingulata* (C, D). In D the tank of *A. lingulata* has been cut open showing the accumulated humus and the root system of a *Clusia* seedling

Since such shoots do not need to support heavy plant biomass, they can afford to have very **wide xylem vessels** reducing friction for the transpiration stream and facilitating transport over long distances (Ewers et al. 1990). Vessels of up to 0.7 mm in diameter have been observed. The xylem sap readily flows out of these wide vessels when they are cut open. Wandering around one may serve oneself from such lianas for a cool drink. Vareschi (1980) reports of a 1 m long piece from which 205 ml of water were collected within 3 min. More detailed studies were performed recently with the tropical vine-like bamboo *Rhipidocladum racemiflorum* (Cochard et al. 1994). Although the xylem vessels are rather wide and efficient for conducting water, they are highly resistant to cavitation which would lead to loss of conductivity. Experimentally, xylem water potentials (see Box **4.1** for explanation of terminology) of -45 bar were required to induce 50 % loss of hydraulic conductivity, but at the end of the 1993 dry season potentials of only – 37.5 bar were reached with a loss of conductivity of 10 % due to cavitation and embolism. In the wet season high root pressures possibly repair cavitation. When transpiration was low in the rainy season, i. e. at night and during rain events positive hydrostatic potentials up to 1.2 bar were built up.

Some species in the aroid genera *Philodendron* and *Monstera* and the Cyclanthaceae *Asplundia* begin their life with rooting in the soil and climb-

Fig. 4.14. Adventitious root-system of the strangler *Clusia rosea*

ing up a phorophyte, but later their old roots degenerate. By growing at the tip and continously degenerating the base of their shoots, they literally crawl up their hosts. Hence, they begin as lianas and later become epiphytes. They have been termed **secondary hemiepiphytes** (Table **4.2: I 1.3.2**). However, this term is not all that convincing; strictly they are secondarily epiphytes. Moreover, in some cases aerial adventitious roots can be formed again, which may hang down from these plants like curtains (Fig. **4.3C**) and establish contact with the ground for a second time. Thus, secondary hemiepiphytes would become **primary hemiepiphytes**, a term used for plants which start their life epiphytically but subsequently establish soil-contact (Table **4.2: I 1.3.1**).

Among the latter are the **stranglers,** a true group of "murderers". Among them is a genus with an extreme plasticity, namely *Clusia* (Clusiaceae, Order Theales) (see also below: Sect. **4.4.1**). *Clusia*-seedlings may get established terrestrially and grow directly as independent trees. However, these plants, like other stranglers, may begin their seedling stage as **humus epiphytes,** using accumulations of humus in knotholes or between branches of phorophytes for establishment (Fig. **4.13 A,** Table **4.2: V 2.1**). Alternatively, they germinate in epiphyte gardens together with several other epiphytic species (Fig. **4.13 B-D**), where tanks of bromeliads or nest and basket-forming ferns provide the required substrate (Table **4.2: V 1.3**). Then, adventitious roots develop, some of which grow positively gravitropically to the ground whilst others are attached to the host tree (Fig. **4.14**). First they may only compete with their host for light. Subsequently, after rooting in the ground, they also compete for nutrients in the soil.

Eventually they strangle and kill their host. Their adventitious roots surrounding the trunk of the host tree hinder the **secondary thickening** and clamp the phloem in the bark with the sieve tubes, the tender pathways for long distance transport of assimilates. Prevented from adequate partitioning of supplies, the phorophyte dies. It seldom falls down. The roots of its ungrateful visitor often form a veritable net with anastomoses via parenchyma bridges, and inside this hollow cylinder of adventitious roots the stem of the former host rots away. Thus, the originally epiphytic strangler becomes an independent tree with a **pseudostem** of adventitious roots. Such behaviour is not only observed by *Clusia,* of course, but equally by other genera with stranglers, e.g. *Ficus* species (Fig. **4.15**). Other species of *Ficus,* i.e. *F. pertusa* and *F. trigonata* live in palm trees. They generate negatively gravitropic roots which suspend them in the crowns of the palms, where they find humus between the leaf bases (Putz and Holbrook 1986). A most successful form is represented by *Ficus bengalensis.* The original aerial roots of a single plant, after gaining ground-contact, may form an entire forest of pseudostems. Walter and Breckle (1984) described an individual, which was only 26 m tall but had an average crown diameter of 170 m, a crown circumference of 530 m and a crown area of 22000 m^2. (See also *Ficus microcarpa* in Fig. **4.15** C.)

Fig. 4.15A-C. Network of anastomosing strangler roots of *Ficus* sp. (**A**) and interior of the hollow cylinder marking the stem of the original host (**B**) in a cloud forest above Lake Coté, Costa Rica. "Forest" of pseudostems of one individual tree of *Ficus microcarpa* (Foster Botanical Garden, Honolulu, Hawaii) (**C**)

4.3.3
Parasites: Mistletoes

Mistletoes growing on bushes and trees are not literally epiphytes, which originally use the phorophytes only as a holdfast. Mistletoes are true parasites. They largely belong to two families of the Order Santalales, namely the Loranthacacae ($\sim$ 900 species and 65 genera) and the Viscaceae ($\sim$ 400 species). Mistletoes occur ubiquitously in the temperate zone, in arid regions as well as in the wet tropics (Sallé et al. 1993). They should be briefly mentioned here because the majority of mistletoe taxa occur in the tropics. Although ecophysiology of mitletoes is increasingly well studied, appartently it is not known why they have such a particularly high diversity and biomass in the tropics (Benzing 1990).

When germinating on the host trees, haustoria of mistletoes penetrate through the bark and join the host cambium, where they form a cambium themselves, which keeps pace with that generating the secondary thickening of the host so that the haustoria gradually become incorporated in the host's wood (Sallé et al. 1993). Via the haustoria the mistletoes establish vascular contacts with the host. Very few mistletoes have phloem connections, since the contacts are predominantly apoplastic between the xylem elements of host and parasite. Thus, the standard view is that mistletoes are hemiparasites on the xylem and transpiration-stream taking only water and nutrients from the host, while they are photosynthetically competent and capable of their own assimilation. The idea that mistletoes might have evolved from terrestrial root hemiparasites sucking the xylem of host roots has recently been discussed (Benzing 1990).

In order to direct part of the transpiration stream from the host to their own shoot system for water and nutrient supply, mistletoes need to establish the required driving force. Indeed, it has been shown that they have a more negative leaf-water potential (see Box **4.1**) and a larger leaf-conductance for water vapour and hence a higher transpiration rate, than the host leaves (Schulze et al. 1984). The difference between the leaf conductances in mistletoes and in their hosts respectively, can also be demonstrated by carbon-isotope analysis (Sect. **2.5**), because in C_3-plants the variable rate of CO_2-diffusion via stomata primarily determines overall changes in ^{13}C-discrimination during photosynthesis. Generally with a small conductance, i.e. when stomata are more tightly closed, if photosynthetic rate is maintained internal CO_2-concentration, $p^i_{CO_2}$, tends to be low. With high conductance, i.e. when stomata are more opened, $p^i_{CO_2}$ approaches the value of external CO_2-concentration, $p^o_{CO_2}$. Overall carbon isotope discrimination, Δ, is then proportional to the ratio of $p^i_{CO_2}/p^o_{CO_2}$ and, hence by indirect association also to leaf-conductance for water vapour, as follows

$$\Delta = a \frac{p^o_{CO_2} - p^i_{CO_2}}{p^o_{CO_2}} + b \frac{p^i_{CO_2}}{p^o_{CO_2}} = a + (b-a) \frac{p^i_{CO_2}}{p^o_{CO_2}} \ [\permil]. \qquad (4.1)$$

In this relationship a gives ^{13}C discrimination due to CO_2-diffusion in air (4.4‰) and b the net fractionation caused by the carboxylation itself (ca. 27‰, i.e. b-a is ca. 22.6‰) so that equation (4.1) becomes

$$\Delta = 4.4 + 22.6 \ (p^i_{CO_2} / p^o_{CO_2}) \ [‰]. \tag{4.2}$$

From the measurements of the carbon isotope ratios in dry plant material, δ_p [see Sect. **2.5**, Eq. (2.2)] and in air, δ_a, Δ is calculated as follows

$$\Delta = \frac{\delta_a - \delta_p}{1000 + \delta_p} \ x \ 1000 \ [‰], \tag{4.3}$$

where δ_a and δ_p are given in ‰. Determined by the contribution from respiratory CO_2 the value for ambient air, δ_a, can vary between -10.5 and -7.5 ‰ especially inside forests and depending on the height above ground. For normal bulk-air conditions, one assumes a δ_a of -8 ‰ (Farquhar et al. 1989a,b; Broadmeadow et al. 1992). The expectation that leaf-conductance for water vapour is higher in the mistletoes as compared to their hosts is in fact borne out by Δ-values obtained (Richter et al. 1995, see also Table **4.4**).

If mistletoes have similar or lower CO_2-assimilation rates as compared to host leaves, this also implies that the mistletoes may have considerably lower water-use-efficiencies (WUE = CO_2 assimilated : H_2O transpired) at the expense of the host. Mistletoes may even grow on mangrove associates like *Conocarpus erectus* (Orozco et al. 1990; see Chap. **5.4**) and true man-

Table 4.4. Overall carbon isotope discrimination Δ (‰) and carbon isotope ratio of dry plant material ($\delta^{13}C$) in the leaves of the mistletoe *Phthirusa ovata* Eichl. (Loranthaceae) and its hosts in the cerrados near Brasília (Reserva Ecologica do Instituto Brasileiro de Geografia e Estatística), Brazil. In the cases labelled by an asterisk (*) one individual of mistletoe was parasitizing three different hosts at the same time. (H. Ziegler, U. Lüttge and A. C. Franco, unpubl. data)

Host species	Carbon isotope ratios in the host		Carbon isotope ratios in *P. ovata*		Δ of mistletoe minus Δ of host
	$\delta^{13}C$	Δ	$\delta^{13}C$	Δ	
Leandra lacunosa Cogn. Melastomataceae	-27.30	19.8	-30.67 -31.09*	23.4 23.8*	3.6 4.0*
Miconia fallax D.C. Melastomataceae	-28.50	21.1	-31.81 -31.09*	24.6 23.8*	3.5 2.7*
Qualea multiflora Mart. Vochysiaceae	-28.31	20.9	-31.09*	23.8*	2.9*
Roupala montana Aubl. Proteaceae	-29.20	21.8	-31.52	24.3	2.5
Miconia albicans (Sw.) Triana Melastomataceae	-29.26	21.9	-30.85	23.6	1.7

groves, where they must establish a water potential gradient large enough to allow movement of water downhill from the salt-loaded halophilic host to their own leaves.

The view, that mistletoes exclusively are parasites for water and nutrients, needs to be modified since carbon gain of mistletoes from the host can be significant (Richter et al. 1995). Studies of partitioning of dry matter and mineral nutrients (Pate et al. 1991a,b), which included analyses of carbon-isotope ratios (see Sect. **2.5**; Marshall and Ehleringer 1990), showed that 24 % (Pate et al. 1991a) to 62 % (Marshall and Ehleringer 1990) of the mistletoe carbon may be derived from the host. This is partially due to the fact, that under various circumstances the xylem sap itself may also carry organic compounds. Secondly the mistletoe tissue may take up organic material from the host phloem by phoem unloading via apoplastic pathways and active membrane-transport, where the mistletoe becomes a sink for source substrates from the host. The involvement of active, membrane-controlled transport can make acquisition of both mineral ions and organic compounds by the mistletoe from the host a highly selective process (Pate et al. 1991b; Rey et al. 1991). Recently, even a heterotrophic fully holoparasitic mistletoe has been discovered, which grows on the tissue of cactus stems (H. Ziegler, pers. comm.).

An interesting morphological feature of mistletoes, related to parasite-host nutrient relations, is the strong resemblance between parasite and host leaves (Ehleringer et al. 1986). The mimicry of host leaves is common since mistletoes often have higher nitrogen contents than their hosts, and hence reduces the likeliehood of mistletoe herbivory. Mimicry is absent when mistletoes are poorer in N than the host.

4.4
Stressors Driving Ecophysiological Adaptation of Epiphytes and Hemiepiphytes

The major factors, which limit epiphytic life and thus may become **stressors** (Box **3.1**), are

- water,
- mineral nutrients and
- light.

Water and nutrients are particularly difficult resources to obtain by epiphytes having no roots in the soil. Light is highly variable, similar to the situation of other species operating in different strata of the forest canopy (see Sect. **3.6**). As already discussed in the context of photosynthesis under dry semi-arid conditions (Sect. **3.7.2**), the effects of these stressors are interconnected in a network of stress interactions, so that adaptive traits are often responses to more than one of them.

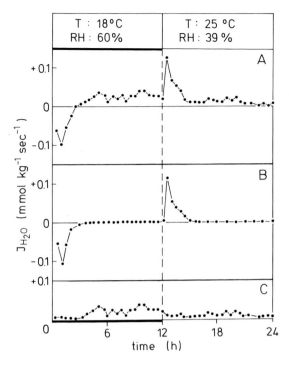

Fig. 4.16 A-C.
Night-day cycle of water-vapour exchange by plants of *Tillandsia recurvata* L. **A** Water-vapour exchange of normal living plants shows a peak of net uptake (negative values of J_{H_2O}) as the dew-point temperature (T) decreases and relative air humidity (RH) increases at the onset of the dark period, and a peak of net release (positive values of J_{H_2O}) with the opposite changes of T and RH at the beginning of the light period. **B** These peaks are also observed with plants killed in boiling water. They are restricted to passive hygroscopic equilibration of dead structures. **C** Subtraction of J_{H_2O} by the dead plants from that of the living plants shows true transpirational water-vapour loss, which is much higher throughout most of the dark period in this CAM bromeliad than during the light period. (Schmitt et al. 1989, from Lüttge 1989)

4.4.1
Water

Rada and Jaimez (1992) compared terrestrial and epiphytic plants of the facultatively epiphytic Araceae *Anthurium bredmeyeri* growing close to each other in a tropical andean cloud forest. The epiphytic plants were affected to a greater degree by the decrease in water availability during the dry season. They showed a larger decrease in leaf conductance and lower leaf water potentials during the dry season than the terrestrially growing plants as well as a reduction in stomatal densities in new leaf growth. Clearly, the water factor can have a large influence on life form of epiphytes.

Most of the lower plant epiphytes, i.e. aerial algae, lichens, bryophytes and even some ferns, are **poikilohydrous** and desiccation tolerant (Table **4.2: III 1,** see also Sects. **4.2; 8.2.2; 8.3.2**). They are truely resistant to drought stress, because they can dry out without suffering damage, overcoming drought periods in a non-hydrated state and becoming viable again when water can be absorbed from precipitation. Of the lichens only those, which have green algae as the photoautothrophic symbionts, are able to acquire their water and reactivate photosynthesis from the water vapour in the gas

Fig. 4.17. Small frog in a tank of a flowering plant of *Bromelia humilis*

phase (Lange et al. 1986, 1988). This also holds for pleurococcoid aerial green algae (Bertsch 1966). However, lichens having cyanobacteria as symbionts require water in liquid form to reactivate photosynthesis.

Some atmospheric bromeliads may take up water from the gas phase of the atmosphere by equilibration of the hygroscopic cell walls of the dead scale cells in the trichomes which densely cover their surface. Thus, one may observe a peak of water-vapour uptake when the relative air humidity (RH) increases at the beginning of the night (Fig. **4.16**). However, this is matched again by a loss of water vapour at the beginning of the day when RH decreased and therefore the bromeliad leaf cells do not have a net gain of water from this mechanism (Schmitt et al. 1989).

In consequence, angiosperm epiphytes have developed a range of other adaptations which often are equally related to the nutrient "stress" factor, e.g. formation of **tanks** or humus collecting **baskets**, in which they effectively create their own soil with a limited water storage capacity. Water demanding animals like small frogs may even live in tanks of bromeliads (Fig. **4.17**), which in some species can impound 5–10 litres of water. Water storage tissues in leaves and stems may also be prominent, so that **leaf and stem succulence** occurs in most bromeliads, orchids and the epiphytic cacti (Fig. **4.18**).

The pre-eminent role of water in limiting the life of epiphytes has resulted in the frequent occurrence of CAM, the mode of photosynthesis which conserves water (see Sect. **3.7.3**). Some authorities have counted

Fig. 4.18A-C. Stem succulent epiphytic cacti (**A** *Epiphyllum*, **B** *Selenicereus inermis* growing through a termite nest), and adaxial water storage tissue of a leaf succulent bromeliad (**C**)

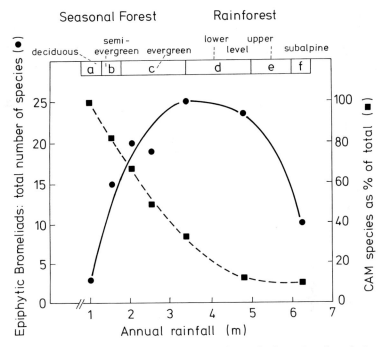

Fig. 4.19. Relations between total number of epiphytic bromeliad species, the relative number of CAM species among them, annual rainfall and prevailing forest types in Trinidad. (Smith 1989)

about 13 500 species of **epiphytes with CAM**. This corresponds to 57 % of all epiphyte species, while only 10 % of all vascular plants are CAM species. The advantage of CAM for epiphytic life is

- water saving, i.e. a high **water use efficiency,**
- provision of an **osmotic driving force** for water uptake by nocturnal acid accumulation,
- **flexibility** in the mode of carbon acquisition

(see Sect. **3.7.3**).

A census of epiphytic bromeliad species in Trinidad has related the frequency of bromeliad epiphytes and the relative number of CAM species to **annual rainfall** and the prevailing type of forest (Fig. **4.19**). Very dry deciduous seasonal forest sustains low epiphytic bromeliad biomass and the small number of species are CAM plants. The abundance of species is highest in the evergreen seasonal forest and the lower montane rainforest. However, the relative contribution of CAM species to the total number of species declines rapidly as forests get wetter and the water saving function of CAM becomes less important.

Table 4.5. Water relation parameters of epiphytes as compared to terrestrial plants. (Simplified from Table 3 of Lüttge 1985). Values were obtained by various authors with different methods as the Scholander pressure chamber technique, plasmolysis measurements and cryoscopy

Plants	Relative water content[a]	ψ	P	π
			(bar)	
Epiphytic CAM plants (ferns, orchids, bromeliads)	0.80 to 0.94	−0.8 to −9.9	2.0 to 10.0	1.7 to 23.4
Terrestrial CAM plants	0.93 to 0.96	−2.4 to −7.9	1.4 to 5.4	4.8 to 12.0
Epiphytic C_3 bromeliad	–	−2.0 to −3.8	2.3 to 3.9	5.2 to 6.1
Terrestrial crop plants	0.58 to 0.75	–	–	–
North American trees	0.60 to 0.85	–	–	–

[a] Ratio of the volume of intracellular water (i.e. inside the plasmalemma) at incipient plasmolysis to the volume of intracellular water at maximum turgor pressure.

Table **4.5** summarizes some **water-relation parameters** (Box **4.1**) of epiphytes. Since most of those studied to date are also CAM-species, it is difficult to decide whether these are typical properties of epiphytes or general characteristics of CAM plants. The **high relative water content** of epiphytes as compared to various C_3-crop plants and trees is noteworthy. C_3-epiphytes, however, have not been studied with this respect. A high relative water content is a typical feature of CAM-plants and is associated with high water-storage capacity and succulence (Fig. **4.20**). For the epiphytic ferns and orchids of Australia Winter et al. (1983) could demonstrate correlations between succulence and CAM expression (Fig. **4.20**). The highest **osmotic pressures** (π) of epiphytes in Table **4.5** are somewhat above 20 bar; in C_3-desert plants they may reach 100 bar. The lowest, i.e. most negative, **water potentials** (ψ) of epiphytes are at -10 bar, in C_3-desert plants values below -150 bar may be found. Hence, the cell sap of epiphytes is diluted, i.e. osmotic pressures are relatively low, the water potential is high and the turgor pressure (P) is low. In this respect epiphytic C_3- and CAM-bromeliads are little different, and also terrestrial CAM-plants show values in this range. The epiphytic C_3/CAM intermediate *Clusia uvitana* (see below for more details on transitions between C_3-photosynthesis and CAM) in a rainforest in Panama has leaf water potentials in the same range, i.e. -7 to -9 bar (Zotz et al. 1994). It was also calculated that stems supported a high leaf area per unit of hydraulic conductivity. In contrast to the vine *R. racemiflorum* (Sect. **4.3.2**), however, *C. uvitana* was very vulnerable to cavitation loosing 50 % of hydraulic conductivity at a stem water potential of only -13 bar (as compared to -45 bar in *R. racemiflorum*). It is possibe that CAM is important in helping to prevent additional damage by insuring low trans-

Fig. 4.20.
Correlation between $\delta^{13}C$ values as a yardstick for CAM expression (see Sect. **2.5**) and leaf thickness indicative of the degree of succulence for Australian epiphytic ferns and orchids. (Winter et al. 1983)

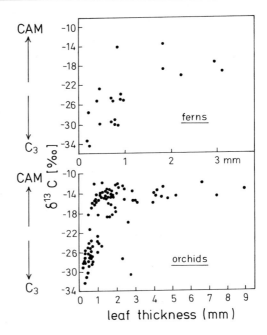

piration rates and avoiding dehydration. Indeed, in comparison with two other epiphytes with different adaptive strategies, *C. uvitana* performed equally well in terms of long-term carbon gain. The comparison (Zotz and Winter 1994b) included

- the evergreen C_3/CAM intermediate *C. uvitana*,
- the drought-deciduous C_3-orchid *Catasetum viridiflavum*, and
- the evergreen C_3-fern *Polypodium crassifolium*.

In *C. uvitana*, during the 4 month dry season, mean daily carbon gain was reduced by ca. 40% following the shift from C_3-photosynthesis to CAM, with strongly decreased daytime CO_2-uptake. *C. viridiflavum* grew new leaves in the second half of the dry season with greatly reduced carbon gain. In the wet season rates of CO_2-uptake by these leaves doubled. Growth occurred until the end of the wet season, when leaves were lost again. In *P. crassifolium* the daily carbon balance was negative during the dry season, and the epiphyte showed characteristics of chronic photoinhibition. Nevertheless, in all of the 3 species there were similar rates of annual carbon gain (1000 g CO_2 m^{-2} yr^{-1}) and long-term nitrogen-use efficiency (i.e. annual carbon gain/mean leaf N content was around 1.1 g CO_2 kg N^{-1} yr^{-1}). The long-term water use efficiency (WUE) of net CO_2 uptake in *C. uvitana* was more than twice that in the other two species.

In CAM plants water relation parameters ψ, P and π also oscillate together with the day-night malic acid rhythm. Fig. **4.21** describes an exper-

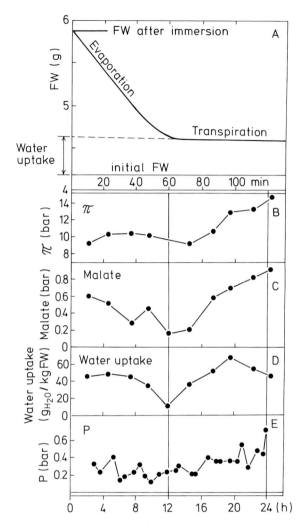

Fig. 4.21A-E.
Experiment showing the
capacity of cells of the
atmospheric CAM brome-
liad *Tillandsia usneoides*
for water uptake during a
day-night cycle. Plants
were weighed (initial FW)
dipped for 10 min into
water, dried superficially
and then weighed at intervals to determine the
point where rapid evapo-
ration of surface water is
completed and water is
only lost from the living
cells by transpiration,
which allows to estimate
water uptake by extrapola-
tion (**A**). It is seen that
osmotic pressure (π, **B**)
and malate levels (**C**)
increase during the night,
and increased water
uptake (**D**) and turgor
pressure (**E**) measured
directly with an intracellular pressure probe are
associated with this. (see
Lüttge 1987)

iment with the atmospheric CAM-bromeliad *Tillandsia usneoides*, showing
that nocturnal accumulation of malic acid provides an **osmoticum**, which
may drive cellular water uptake. It can be seen that cell-sap osmotic pressure
(π) increases together with malic-acid levels. Water uptake, measured after
dipping the plants for a short period into water as shown in Fig. **4.21 A**,
clearly increased during the night together with π, and this also led to an
increase in turgor pressure. It should be noted, that while atmospheric bromeliads could occur in rather dry habitats, they are often found at sites where
fog forms during the later part of the night and in the early morning. Water
from condensed fog and dew is then available at times when malic-acid concentration in the cells is high and can lead to osmotic uptake of water.

A major advantage of CAM in habitats where there are large short term and seasonal variations in water availability is the inherent **flexibility** in mode of carbon acquisition. The **different expression of the four CAM phases** (see Box **3.11**) in constitutive CAM plants already allows highly variable responses. If water supply were to range from severe to moderate and low drought stress, there may be, respectively total stomatal closure and CO_2-recycling, predominant nocturnal opening of stomata (Phase I) or increasing use of phase IV and phase II CO_2-uptake during the daytime hours. There may even be continous CO_2 uptake day and night under well watered conditions. In addition there are species which are true **C_3-CAM**

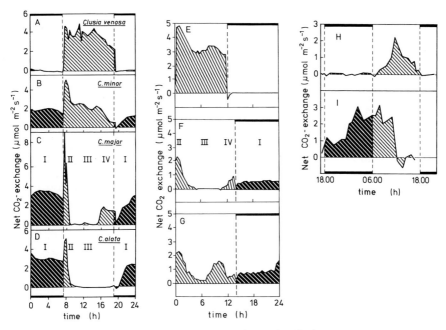

Fig. 4.22 A-I. Modes of photosynthetic CO_2 exchange in *Clusia*
Left panel. Comparison of four species under identical conditions in a phytotron, *Clusia venosa* with C_3 photosynthesis (**A**), *Clusia* minor with CO_2 uptake day and night (**B**), *Clusia major* and *Clusia alata* both with CAM but differing in the development of phase IV (**C** and **D**)
Center panel. *Clusia minor* in a growth chamber with C_3 photosynthesis under well-watered conditions at high irradiance (1700 μmol photons m^{-2} s^{-1}) and medium leaf/air water vapour pressure difference (ΔW = 6.6 mbar bar^{-1}) (**E**); CAM with the well-expressed four phases (*I–IV*) under drought stress at low irradiance (400 μmol photons m^{-2} s^{-1}) and high ΔW (13.5 mbar bar^{-1}) (**F**); and CO_2 uptake day and night under well-watered conditions, low irradiance (400 μmol photons m^{-2} s^{-1}) and low ΔW (3.4 mbar bar^{-1}) (**G**)
Right panel. *Clusia rosea* in the field with C_3 photosynthesis (**H**) and CAM with an extended phase II in the first half of the day (*I*)
Black bars on the abscissa indicate the dark periods. (Lüttge 1991)

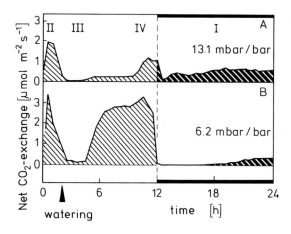

Fig. 4.23A, B.
CO_2 gas exchange of two opposite leaves at the same node of *Clusia minor* in a growth chamber. After 4 days of drought stress the plant was rewatered on the 5th day (*arrow head*). The leaf kept in a cuvette at high VPD (ΔW = 13.1 mbar bar^{-1}) continued to perform CAM (**A**) with all phases *I – IV* expressed. The other leaf (**B**) at low VPD (ΔW = 6.2 mbar bar^{-1}) rapidly switched to daytime CO_2 uptake and suppressed nocturnal CO_2 uptake. The *black bars* on the abscissa indicate the dark period. (Lüttge 1991)

intermediates. They can switch from C_3 photosynthesis to CAM as drought stress increases, and back again when the stress is released. Among the epiphytic bromeliads *Guzmania monostachia* is such a C_3-CAM intermediate. The epiphytic fern *Pyrrosia confluens* and the Crassulaceae *Kalanchoë uniflora* also belong to this group (Griffiths 1989) as well as species of *Peperomia* (Sipes and Ting 1985; Ting et al. 1985; Holthe et al. 1987). The plants showing the most flexible response, however, are in the hemiepiphyte and strangler genus *Clusia*. Each of the photosynthetic modes described above, viz. pure CAM, pure C_3 photosythesis and CO_2-uptake day and night, have been observed with *Clusia* (Fig. **4.22**).

Variability of ecophyiological response is observed

- between different species under given environmental conditions (Fig. **4.22, left panel**)
- for a given species under different environmental conditions (Fig. **4.22, center and right panel**), and even
- for the two different leaves of a given node in the same plant when they are kept under different conditions (Fig. **4.23**).

The rapid changes between C_3-photosynthesis and CAM, that may be performed by *Clusia* are determined by the external control parameters

- i)   **water relations,**
- ii) **day/night temperature regime** and
- iii) **light.**

i)   In an experiment with a plant of *C. minor* rewatered after a period of several days of drought, it was possible to get two opposite leaves at a given node to perform C_3-

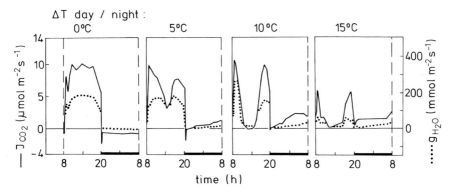

Fig. 4.24. Change of *Clusia minor* from C_3 photosynthesis to CAM as night-time temperatures are lowered to give an increasing day/night temperature difference (ΔT). J_{CO_2} net CO_2 exchange, g_{H2O} leaf conductance for water vapour. Day-time temperature was always 30 °C. (Haag-Kerwer et al. 1992)

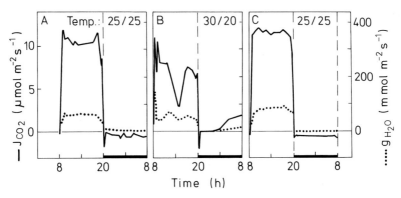

Fig. 4.25A-C. Change of *Clusia minor* from C_3 photosynthesis to CAM and back to C_3 photosynthesis as a day/night temperature difference of 10 °C is introduced and removed again. Gas exchange after 7 days at 25/25 °C day/night (**A**) followed by 5 days at 30/20 °C (**B**) and by 4 days at 25/25 °C (**C**). J_{CO_2} net CO_2-exchange, g_{H2O} leaf conductance for water vapour. (Haag-Kerwer et al. 1992)

photosynthesis and CAM respectively, at the same time. One leaf, when maintained in an atmosphere with a low leaf-air water **vapour pressure difference** (ΔW or VPD), i.e. kept under a low transpiratory demand, switched to C_3-photosynthesis a few hours after watering with CO_2 uptake markedly reduced in the subsequent night. The other leaf, kept at high VPD, continued to perform CAM with the 4 phases clearly noticeable, as both leaves had done during the drought period before watering (Fig. 4.23).

ii) By varying the **temperature regime,** it was found that a certain day-night temperature difference was important for expression of CAM in *C. minor* (Fig. 4.24). The shift between CAM and C_3-photosynthesis was fully reversible when the temperature regimes were changed between equal day/night temperature and day/night temperature differences (Fig. 4.25).

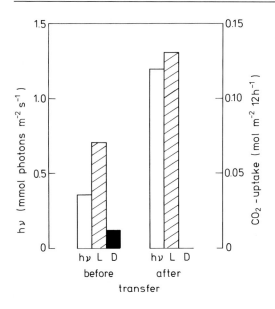

Fig. 4.26.
Elimination of dark CO_2 fixation (*D*) and stimulation of light CO_2 fixation (*L*) in a well-watered plant of *Clusia minor* by transfer from lower to high irradiation (*hv*). (Data from Schmitt et al. 1988)

Table 4.6.
The highest nocturnal acid accumulation (mmol titratable H^+/l) and the highest nocturnal citrate accumulation (mmol citrate/l) observed in CAM. These records were measured in epiphytes (*Aechmea nudicaulis*) and hemiepiphytes (*Clusia* species)

Titratable acidity	
Aechmea nudicaulis	
Field Trinidad, March 1983	625[a]
Clusia rosea, phytotron	1120[c]
Clusia minor	
Field, Trinidad, March 1990	1410[b]
Citrate	
Clusia minor,	
Field Trinidad, March 1990	125[b]
Clusia rosea, phytotron	200[c]

Data from [a] Smith et al. 1986b, [b] Borland et al. 1992, [c] Franco et al. 1992.

iii) In well-watered plants a drastic increase in **light intensity** led to an elimination of nocturnal dark-CO_2-fixation and an increase in daytime C_3-photosynthesis. Obviously, this represents an optimal use of high light energy provided that water is not limiting (Fig. **4.26**).

Clusia spp. are also remarkable in several other ways:
- showing the **highest nocturnal acid accumulation** ever observed for CAM plants (Table **4.5**),
- accumulating large amounts of **citric acid** during the dark period additionally or alternatively to malic acid (Table **4.6**).

The latter observation also requires a comparative evaluation of the relative ecophysiological advantages of malic and citric acid accumulation during CAM (Table **4.7**). Consideration of intermediary metabolism suggests that different compartmentation and differ-

Table 4.7. Comparative evaluation of the ecophysiological advantages of nocturnal accumulation of malate (Δ malate) and citrate (Δ citrate) respectively, in CAM. (Franco et al. 1992; Haag-Kerwer et al. 1992)

	Δ **malate**	Δ **citrate**
Carbon acquisition	Yes	No
H_2O-saving during C acquisition	Yes	No
$\Delta\pi$ with possible H_2O acquisition	Yes	Limited
Nocturnal recycling	CO_2	Carbon skeletons
Daytime recycling	CO_2	CO_2
Buffering capacity	Small	Large
Energy requirements		
Dark period: for hexose breakdown		
and vacuolar organic acid accumulation	– Similar –	
Light period: for recovery	Slightly	Slightly
of hexose	smaller	larger
Ecophysiological functions	H_2O-saving; preventing photoinhibition to some extent	Effectively preventing photoinhibition

ent contributions of mitochondrial and cytosolic reactions may both be involved. Citric acid accumulation, in contrast to malic acid accumulation, does not lead to a net gain of carbon, although it contributes to carbon recycling. However, carbon recycling via citric acid may be favourable because daytime breakdown of citric acid may possibly result in the liberation of more CO_2 than that of malic acid, and the availability of this internal CO_2 could prevent photoinhibition (see Sect. 3.7.3) more effectively when light intensity is high. In fact it has been observed for 4 different *Clusia* species that the ratio of malic acid: citric acid accumulated during the dark period decreased in response to drought stress, relatively favouring carbon recycling via citrate. The energy turnover is not appreciably different in both cases. Since citrate accumulation does not contribute to C-acquisition, naturally it also does not help to improve water-use-efficiency (WUE). It also adds less to changes in cell-sap osmotic pressure than malate accumulation, because only one mole of citrate is formed per mole of hexose consumed. However, citrate is known to be an effective buffering substance. This may sustain the very high nocturnal vacuolar acid accumulation observed in *Clusia* with day-night changes of titratable proton levels of more than 1 M (Table **4.6**).

The diversity of the hemiephyte and strangler *Clusia* in the tropics has presented us with many unexpected observations and stimulating new reflections on the nature of ecophysiological adaptations. There are ca. 145 species of *Clusia* occupying a wide range of habitats, e.g. coastal rocks and sand dunes, gallery forests, savannas, rock outcrops (inselbergs; see Sect. **8.3.1**), low land and upper montane rainforests, cloud and elfin forests (Fig. **4.27**, see also Fig. **3.3 D**). Perhaps it is so successful because of the high degree of physiological plasticity. This also makes it suitable for reclamation of tropical land by afforestation (Fig. **4.28**) and as an ornamental tree even in the center of cities.

Fig. 4.27A-E. Habitat diversity of *Clusia*. **A** *Clusia fluminensis* on sand dunes in the restinga formation on the Atlantic coast near Rio de Janeiro, Brazil. **B, C** *Clusia rosea* on granite rocks on the British Virgin Island Virgin Gorda (Lesser Antilles), with aerial adventitious root systems in the rock furrows in **C**. **D** *Clusia* sp. Gran Sabana, Venezuela, with an epiphytic bromeliad *Catopsis berteroniana*. **E** *Clusia rosea* in montane rainforest on the US Virgin Island St. John (Lesser Antilles).

Another genus of hemiepiphytes and stranglers, which has a habit very similar to that of *Clusia*, namely *Ficus*, has been studied much less in terms of physiological ecology of photosynthesis and water relations. This is astonishing because the genus appears to be as successful in tropical forests as *Clusia*. Both have very different strategies though. As far as it is known to

Fig. 4.27 D, E

date all species of *Ficus* are obligate C_3-plants. Holbrook and Putz (1996b) have made interesting intraspecific comparisons of water relations in the life forms of epiphytes and terrestrial trees in the genera *Ficus* and *Clusia*. They found that in 5 species of *Ficus* the epiphytic life forms as compared to terrestrial trees had

- several-fold higher specific leaf area (m^2g^{-1}), which may also be taken as higher degree of "succulence";
- 2-4fold lower stomatal densities, which may be discussed in relation to the need of reduced transpiration at lower availability of water in the epiphytic habitat;
- osmotic pressures (π) about 6 bars lower,
- and a bulk modulus of cell wall elasticity (ε) about 50 % lower.

We must note that cell wall elasticity is inversely related to ε, i. e. the higher ε the more elastic and the lower ε the stiffer is the cell wall. With the relationship of

$$\Delta P = \varepsilon \cdot \frac{V}{\Delta V}, \tag{4.4}$$

Fig. 4.28.
Afforestation of a steep
savanna plot in the
grounds of the Instituto de
Investigaciones Cientificas
(IVIC) at 1200-1300 m
above sea level, Caracas,
Venezuela. Owing to its
highly adaptive photosyn-
thesis and the effective
root system, *Clusia* is particularly well suited for
reclamation of such
deforested land

where ΔP and ΔV are turgor pressure and volume changes, respectively, and V is cell volume, it then follows that a given change in volume (ΔV) leads to a lower change in pressure when ε is larger or cell walls are stiffer, i.e. in the epiphytic *Ficus* plants a larger volume of water can be lost before turgor is lost than in the terrestrial trees. As a result leaves of the more succulent epiphytes and the conspecific less succulent trees of *Ficus* species lost turgor at approximately the same relative water content.

With the water potential

$$\psi = P - \pi \tag{4.5}$$

(see Box **4.1**) at P = 0, or zero turgor $\psi = -\pi$, and this is substantially higher (less negative) in the epiphytes due to the lower osmotic pressure π. These observations agree with the general trends for higher succulence, higher ψ and lower π in epiphytes (see above and Table **4.5**).

In contrast to *Ficus* species, in *Clusia* differences in all of these water relation parameters between epiphytes and trees were very small. Thus, while

Clusia has instantaneous plasticity of responding to changing water supply and evaporative demand by photosynthetic options (C_3-CAM transitions) *Ficus* shows intrinsic developmental changes during the transformation from epiphyte to tree which is associated with improved acquisition of water.

4.4.2
Mineral Nutrients

Some special adaptations to the poor nutrient supply in the epiphytic habitat have already been mentioned above in the discussion of life forms of epiphytes (Sect. **4.3.1**) and in relation to water stress (Sect. **4.4.1**). They include the use of **humus accumulation** in trees (Fig. **4.13 A**) and morphological features of the plants for collecting humus such as the formation of **baskets** and **nests** (Fig. **4.13 B**) as well as tanks (Fig. **4.11** and **4.13 C, D**). **Scales** (epidermal trichomes) of bromeliads (Fig. **4.10**) and the **velamen radicum** of aerial roots of aroids and orchids serving atmospheric nutrition also belong to the specialised plant structures formed for nutrient and water uptake. The velamen is a multilayered peripheral structure, which is readily infiltrated by water from throughfall or stemflow (Fig. **4.29**).

It has even been argued that the successful trapping of rain and throughfall, enriched by leachates from leaves and stems, is **nutritional piracy**, depriving host trees of resources which otherwise would reach their rooting medium (Benzing 1989a,b, 1990). A comparison of the nitrogen content in leaves of facultative epiphytes at adjacent sites, showed significantly lower N-levels in two aroid species growing epiphytically as compared to their terrestrial counterparts. N-content was similar in two tank-forming bromeliads, whereas in seedlings of *Clusia* low N-content was also related to growth directly on the phorophyte and not to growth inside bromeliad tanks (Fig. **4.30**). Epiphytes may also form mycorrhizas. The fungal hyphae penetrate the epiphytes as well as the decaying bark of the host phorophyte. This is evidently piracy of a more overt kind and the tree effectively becomes the pedosphere of the epiphyte (Ruinen 1953; Johannson 1977; Benzing 1982; Benzing and Atwood 1984). An interesting example of counter-piracy is found when phorophytes produce adventitious canopy roots which exploit the nutrient debris collected within the epiphyte cover (Nadkarni 1981).

Stewart et al. (1995) have used analyses of the stable isotope ^{15}N to trace several possible sources of nitrogen and processes of nitrogen acquisition in tropical epiphytes as compared to associated soil rooted trees. The content of ^{15}N in the epiphytes was comparatively low. This suggests that epiphytes may make considerable use of atmospheric N-deposition and dinitrogen (via biological N_2-fixation), i.e. N-sources depleted in ^{15}N. Cyanobacteria and free living N_2-fixing bacteria of the phyllosphere (see Sect. **4.1**) and other epiphytic habitats may play an important role in supplying N to epiphytes.

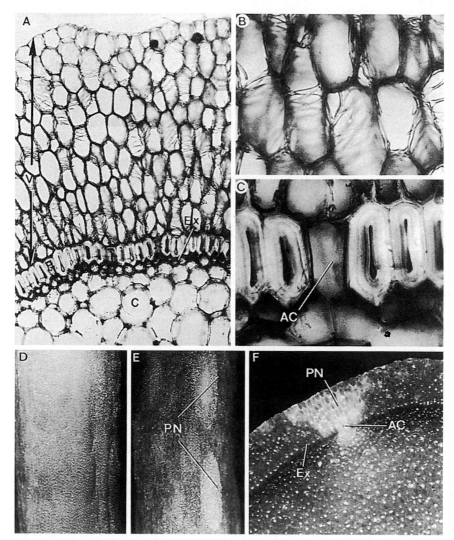

Fig. 4.29A-F. Velamen radicum in epiphytic orchids. **A** Velamen of a *Dendrobium* species. *V* velamen; *Ex* exodermis; *C* cortex. **B** Detail from **A** showing the dead velamen cells
with the typically perforated walls. **C** Detail from **A** showing the exodermis with an aeration cell (*AC*). **D** Surface view of an aerial root of *Vanda tricolor* with dry velamen. The
air-filled velamen cells appear homogeneously whitish. **E** The same detail as in **D**; however, after wetting the velamen. With the exception of the pneumatothodes (*PN*), the velamen cells are filled with water and thus appear dark. **F** Cross-Sect. through the water-
imbibed velamen of *Vanda tricolor* in the region of the pneumatothode (*PN*) and aeration
cells (*AC*). The air-filled cells of the pneumatothode appear white. (Goh and Kluge 1989)

Fig. 4.30.
Comparison of nitrogen levels in cohabitant epiphytic and terrestrial-life-forms of two aroid and two bromeliad species and of *Clusia rosea*. Data show N content in leaves of epiphytic minus leaves of terrestrial plants of the same species (numbers are p values of a t test for statistically significant differences or n.s. = nonsignificant). (Ball et al. 1991)

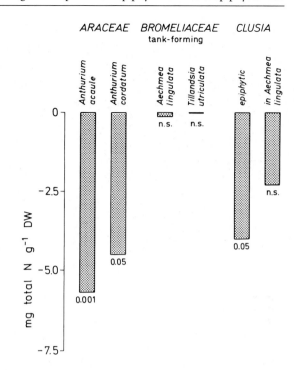

Two additional strategies involve **interactions with animals,** one of which is predatory and the other one symbiotic, namely

– **carnivory** and
– **mutualism with ants.**

Carnivory by plants is quite frequent in the tropics and subtropics (Fig. **4.31**). In the plant kingdom, carnivory is generally assumed to be a mechanism for the acquisition of mineral nutrients, especially N, P and S, by photosynthetically autotrophic plants living in nutrient-poor habitats such as peat bogs (Schmucker and Linnemann 1959; Lüttge 1983). Hence, one would assume that **carnivorous plants** would be rather frequent in the canopy habitat. However, this is not the case. Carnivorous plants have developed special organs for the capture of prey, glands for digestion and absorption of low molecular compounds obtained from the prey, and mechanisms for the attraction of small animals such as showy and colourful appendages and production of scent and nectar (see also Sect. 7.3.3.3). Thus, Givnish et al. (1984) have explained the rarity of carnivorous plants in epiphytic habitats by the high costs for investment and maintenance of these complex attraction and capture mechanisms. Since other factors, particularly water and often light are equally limiting, a cost-benefit analysis suggested carnivory would not be effective under these circumstances.

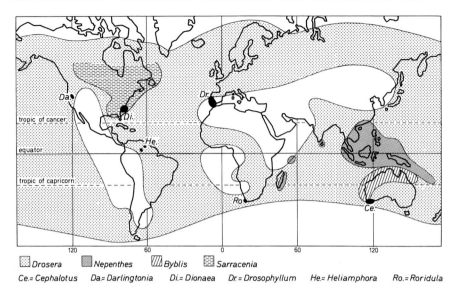

Fig. 4.31. Global distribution of carnivorous plants. The genus *Drosera* is almost ubiquitous and occurs *between the two* lines drawn in the north and in the south respectively, except in the areas left *white*. The exclusively tropical and subtropical distribution of the genera *Nepenthes* and *Byblis* is clearly seen and the islands of endemic *Heliamphora* in Northern South America should be noted. Distribution of carnivorous genera of the family Lentibulariaceae is not shown on this map, namely *Pinguicula:* Northern temperate zone; *Genlisea:* Tropical South America, South Africa; *Utricularia:* cosmopolitan; *Biovularia:* Cuba, Tropical South America; *Polypompholyx:* South America, Australia. (After Schmucker and Linnemann 1959)

On the other hand, there are a few examples of carnivorous plants among climbers and epiphytes. The pitcher plant genus of *Nepenthes* is native in the Malaian archipelago and an exclusively tropical genus (Fig. **4.31**). There are 71 species of *Nepenthes* altogether, among which a few are purely terrestrial, but 6 are epiphytic and many are climbers (Fig. **4.32**). The pitchers attract prey by their shiny and often colourful rim, which also bears nectaries towards the inside of the pitcher opening. Small animals, predominantly insects, having fallen over the slippery collar into the pitcher lumen, rapidly drown in the digestive fluid produced by glands on the bottom, which contains a protease secreted by the plant and other enzymes provided by microorganisms participating in prey digestion. Escape via the pitcher walls is prevented by downward pointing scale-like tissue over the glands (**Fig. 4.3.3. C**), modified stomata with protrusions towards the pitcher lumen and a lubrication with small and loose wax particles in the upper part of the pitcher (Fig. **4.33 A, B**). Substances obtained from the digested prey are absorbed via the gland cells.

Some of these traits are also shared by tanks of bromeliads. They often contain dead and putrefying insects and may absorb substances like amino acids from such prey via their scales. In some cases, such as the epiphytic bromeliad *Catopsis berteroniana*, there is also wax at the adaxial leaf surfaces lubricating the tank interior (Benzing 1989b). However, these plants have no glands and do not secrete digestive enzymes, so that at most there is only the initial development towards the carnivorous syndrome (see also Sect. **7.3.3.3** below).

Fig. 4.32
Nepenthes gracilis
(Malaysia)

A more sophisticated example is *Utricularia*. Many species in this genus are aquatic, forming small bladders from modified leaves along the stems. The bladders are tightly closed by a trap-door, and actively transport ions across the trap wall into the outer medium to drive the osmotic loss of water from the trap lumen. This creates tension in the bladder-walls, which sets the bladders. Small animals trigger the opening of the trap door by touching the antennae-like protuberances and are swept into the trap as the tension in the trap wall is released (Fig. **4.34**). The animals are then digested inside the bladders. In the tropics, *Utricularia* species often live epiphytically between mosses on stems of trees. Most cunning is *Utricularia humboldtii*, which lives inside the tanks of the bromeliad *Brocchinia tatei* (Fig. **4.35**).

Another tropical habitat which is often nutrient limited are the savannas. We will refer to carnivorous plants again below, when we discuss their role and their contribution to nutrient turnover in this habitat (Sect. **7.3.3.3**).

In **symbiotic mutualisms of epiphytes with ants** we may distinguish two forms, which among other benefits, provide mineral nutrition to plants, namely

- **ant garden epiphytes** and
- **ant house epiphytes**

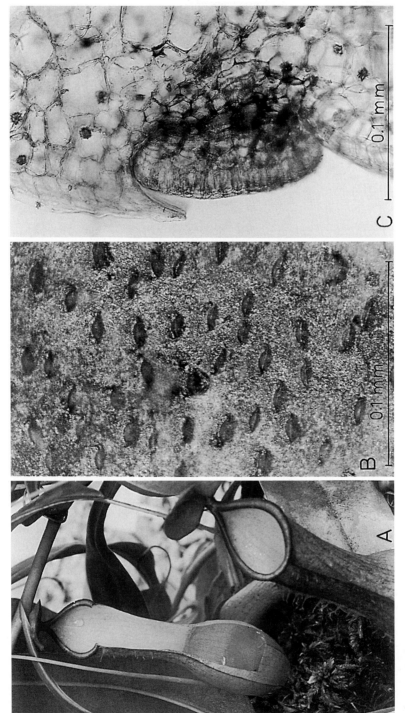

Fig. 4.33A–C

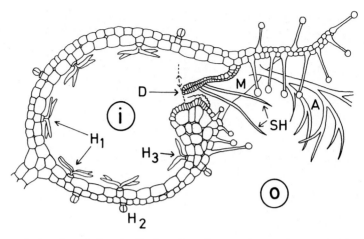

Fig. 4.34. Schematic drawing of a longitudinal section of a trap of *Utricularia*. *i* Bladder lumen; *o* outer medium; *D* trap door; *M* trap mouth; *SH* sensitive hairs of the trap door = trigger hairs; *A*, antenna. *Dotted arrow* opening and closing of the trap door; the tissue beneath the trap door prevents opening from the inside. H_1, H_2, H_3 various types of gland hairs serving prey digestion and transport functions. (After Schmucker and Linnemann 1959).

(Davidson and Epstein 1989; Benzing 1989b, 1990). Ants frequently construct nests in trees using various materials which are rich in nutrients (forming an **ant-nest "carton"**). Seeds of plants may germinate directly from such a nutritive carton. Since the plants offer various goods in return, such as nectar, fruits and seeds, the ants often disperse and plant the seeds of their epiphytes in ant gardens. Conversely, plants themselves may also provide nesting facilities for ants, e.g. cavities in various parts of the plant body or hollow stems (Figs.. **3.24** and **4.36**; Sect. **3.4.3.2**). Among the epiphytes there are many **myrmecophytic** species with such ant houses, e.g. orchids with ant nests in pseudobulbs and bromeliads with inflated tank leaves (Fig. **4.36A – C**). The ants carry soil and other decaying material into the nests and add their faeces, which gives a debris from which the host plants may absorb nutrients. The pitchers of some species of the Asclepiadaceae *Dischidia* are most sophisticated, since adventitious roots grow into the soil and debris accumulated by ants inside these containers. In effect, epiphytic *Dischidias* literally construct their own flower pots (Fig. **4.36D**). A recent study with the Malaysian *Dischichia major* using stable isotopes to trace sources of N and C (Sect. **2.5**) has shown that 29 % of the host nitrogen is derived from debris deposited into the leaf pitchers by ants and that 39 %

◀**Fig. 4.33A-C.** *Nepenthes.* **A** Pitcher cut open showing the velvety covering with loose wax particles in the upper part and the gland zone below. **B** Part of the upper pitcher region with wax particles and protrusions. **C** Scale-covered gland

Fig. 4.35A-C. *Utricularia humboldtii* in the tanks of *Brocchinia tatei*. **A** Inflorescence of *U. humboldtii* emerging from a tank of *B. tatei*. **B** Outer tank leaves removed to show the basal system of *U. humboldtii* bladders on stems and leaves with petioles and lamina emerging from the tank. **C** Larger and smaller *U. humboldtii*-bladders

Fig. 4.36A-D. Ant-house epiphytes. **A, B** The orchid *Schomburgkia humboldtiana* with pseudobulbs cut open in **B** to show the ants nest. **C** The bromeliad *Tillandsia flexuosa* epiphytic on the cactus *Pilosocereus ottonis*, where ants are nesting in the inflated basal part of the tank. **D** The Asclepiadeaceae *Dischidia rafflesiana* with adventitious roots in modified leaves

of the carbon assimilated by the host is derived from ant-related respiration (Treseder et al. 1995). Ant respiration may increase the CO_2 concentration in the pitcher lumen above atmospheric levels. Since the host has stomata on the inner pitcher-wall surface this can be directly used for CO_2-fixation even in the dark in this obligate CAM-plant.

In many cases these plant-ant interactions are true symbioses with obligate mutualism, since the partners are no longer successful individually. The epiphytic plants benefit nutritionally and may be protected from herbivores, while the ants obtain nest-sites and various items of food.

4.4.3
Light and the Evolution of Plants to Epiphytism

Most textbooks still suggest that climbing and epiphytism in plants is a **struggle for light** in an escape from the darkness of forest floors. This goes back to A.F.W. Schimper (1888), who concluded his observations in forests of the American tropics with the hypothesis that epiphytic bromeliads evolved from shade adapted terrestrial forms. However, as we have already mentioned above (Sect. **4.1**), lianas and vines most frequently are light demanding plants of pioneer successions.

Epiphytes occupy sites of variable light exposure. Studies of the distribution of epiphytic orchids on phorophytes in a West African rainforest have shown that only a small percentage of the total number of species are found in the upper canopy, and most species dwell within the crowns of trees (Fig. **4.37**).

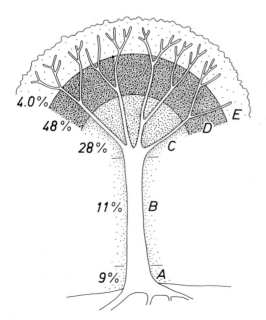

Fig. 4.37.
Distribution of epiphytic orchids on trees in a West African rainforest. Numbers of species found in the different zones of the phorophyte related to total orchid species counted. The zones of the phorophyte are: *A* the basal part of the stem up to 3 m above ground level; *B* the stem up to the first ramifications; *C, D* and *E* the canopy divided into three equal parts along the length of the branches from inside to outside. (Goh and Kluge 1989, after Johansson 1975)

Fig. 4.38.
Distribution of epiphytic
Bromeliaceae of the expo-
sure group (*Ex*), the sun
group (*Su*) and the shade-
tolerant group (*Sh*) with
C_3 photosynthesis (*open
parts of the bars*) and
CAM (*closed parts of the
bars*) in Trinidad related
to annual rainfall. (After
Griffiths and Smith 1983)

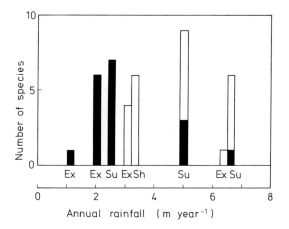

Pittendrigh (1948) grouped the **epiphytic bromeliads** of Trinidad in 3
categories according to their light demand:

- an **exposure group,**
- a **sun group** and
- a **shade tolerant** group.

Using stable carbon isotope analysis (Sect. **2.5**) Griffiths and Smith (1983)
have determined the distribution of C_3-photosynthesis and CAM among
these 40 species. They related the mode of photosynthesis to Pittendrigh's
light-demanding categories and the annual precipitation at the sites where
they occur in Trinidad. The result of the survey is depicted in Fig. **4.38** (see
also Fig. **4.19**). At the wettest site ($>$ 6.4 m precipitation per year) the shade
group is not represented at all, with one species of the exposure group, 6
species of the sun group and only 1 CAM species being present. At somewhat lower annual precipitation, the sun group prevails with a total of 8 species and 3 CAM species among them. Under intermediate precipitation,
only C_3 species comprise the exposure and shade groups. However, at the
driest sites one finds only CAM plants of the exposure and sun groups.
Thus, CAM among epiphytic bromeliads is clearly correlated with reduced
water availability and sun exposure which exacerbates drought stress.

Together with the development and specialization of epidermal tri-
chomes (see Sect. **4.3.1.**, Table **4.3**, Fig. **4.10**), which can be considered as an
evolutionary trait (Mez 1904; Tietze 1906), Pittendrigh (1948) used the
abundance and distribution of species in the 3 categories for consideration
of the evolution of epiphytism among bromeliads. He suggested that epiphytic bromeliads did not evolve from shade demanding ancestors of the
forest floor but rather were derived from terrestrial ancestors of open habitats originally adapted to sun exposure and at least temporary drought.

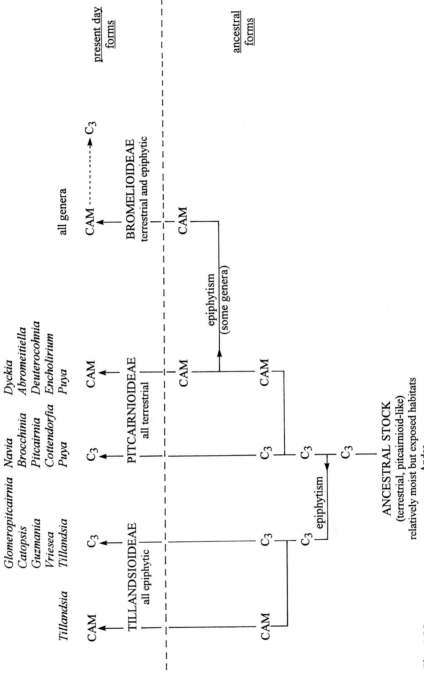

Fig. 4.39

The observation that CAM occurs only among bromeliads of the sun and exposure groups supports this interpretation because phylogenetically CAM is considered to be rather an advanced physiological trait. Thus, Smith (1989) developed the scheme of possible **phylogenetic relationships within the Bromeliacae** shown in Fig. 4.39. From a pitcairnioid-like ancestral stock of terrestrial plants (see also Sect. **3.7.2**) epiphytism and CAM must have evolved in the bromeliads independently several times. In conclusion, at least for the bromeliads, Schimper's idea of epiphytic evolution towards light from shade demanding understory plants is most probably not valid. It was more likely to have represented a conquest of space by plants already adapted to the ecophysiological problems of exposed nutrient-poor habitats. The shade-adapted epiphytic bromeliads are mostly shade-tolerant and not shade-demanding, and they probably constitute a later development.

However, Schimper's hypothesis may still apply to the **evolution of epiphytism** in other groups of plants. Where this is the case, one would expect to find not only shade-tolerant but also clearly shade-demanding species. This can be assessed by comparing the cardinal points of light-dependence curves, which distinguish shade and sun plants (see Sect. **3.6.2.1**) (Table **4.8**). By the criteria of light-compensation point, light-saturation and maximal rates of photosynthesis at least some epiphytic ferns and orchids are found to be typical shade plants, while other ferns and orchids, *Kalanchoë uniflora* and *Nepenthes* appear to be sun types.

The fossil record of epiphytes (Mägdefrau 1956) is meagre and does not offer support for the discussion of evolution. The upper Devonian arborescent horsetail, *Pseudoborina ursina* about 360 million years ago had a stem diameter of 0.12 m and carried an epiphyte *Codonophytum epiphytum*. The phylogenetic relationships, however, are uncertain. In Permian formations in Germany from about 260 million years ago leafy organs resembling basket forming mantle leaves of extant epiphytic ferns of the genera *Platycerium* and *Polypodium* are preserved. However, these extant fern genera are leptosporangiate, and such forms did not exist at that time. The most original extant pteridophytes are two species each of the genera *Psilotum* and *Tmesipteris* and all of them are epiphytes. They belong to the Psilotatae and are often considered as relicts of evolution. Although they are very similar to the earliest land plants, i.e. the Psilophytatae, which first conquered land and died out again in the middle Devonian 360 million years ago, the exact phylogenetic relationships again are not clear. From the very poor palaeontological record alone one may readily conclude that epiphytism of vascular plants is a very recent event in geological history.

◀**Fig. 4.39.** Scheme of possible phylogenetic relationships within the Bromeliaceae based on the taxonomic distribution of CAM and C_3 photosynthesis at the level of individual genera. The scheme shows that both the epiphytic habit and CAM must have arisen more than once during evolution of the present-day forms. Within the Bromelioideae there are indications of a progressive loss of CAM in some genera. (Smith 1989)

Table 4.8. Cardinal points of light-response curves of various epiphytes compared to genuine sun and shade plants. Epiphytes were related to sun and shade plants respectively, by evaluating all of the three given criteria (light-compensation point, light-saturation of photosynthesis, rate of photosynthesis at saturation) together, because coordination is not simple when using single criteria individually. (After Lüttge 1985)

Plant type or species	Light compensation point (μmol photons m^{-2} s^{-1})	Light saturation of photosynthesis (μmol photons m^{-2} s^{-1})	Rate of photosynthesis at saturation (μmol CO_2 or O_2 m^{-2} s^{-1})
Sun plants[a]	20–30	400–600	10–20
Epiphytes:			
Platycerium grande (C_3 fern)[b]	20	$\gg$ 520	> 3
Anthurium hookeri (C_3 aroid)[b]	5–40	180–375	1–6
Kalanchoë uniflora (C_3/CAM	25–75	275–500	0.5–4
Crassulaceae)[b]	16–20	180–200	4–8
Phalaenopsis violacea (CAM orchid)[c]	14–20	240–260	6–8
Phalaenopsis grandifolia (CAM orchid)[c]			
Shade plants[a]	0.5–10	60–200	1–3
Epiphytes:			
Aglaomorpha heracleum (C_3 fern)[b]	5	160	3
Pyrrossia longifolia (CAM fern)[b]	8	100–150	1–2
Pyrrosia lanceolata (C_3 fern)[b]	15	200–300	–
Drymoglossum piloselloides (CAM fern)[b]	8	300–500	–
Nepenthes × *hookeriana* (C_3 carnivorous plant)[b]	5-10	150–225	3–5

[a] Data as given also in Table **3.3**.
[b] Data for glass-house grown plants (Lüttge et al. 1986).
[c] Data for field-plants in Singapore (Goh and Kluge 1989).

References

Badger MR, Pfanz H, Büdel B, Heber U, Lange O (1993) Evidence for the functioning of photosynthetic CO_2-concentrating mechanisms in lichens containing green algal and cyanobacterial photobionts. Planta 191:57–70

Ball E, Hann J, Kluge M, Lee HSJ, Lüttge U, Orthen B, Popp M, Schmitt A, Ting IP (1991) Ecophysiological comportment of the tropical CAM-tree *Clusia* in the field. I. Growth of *Clusia rosea* Jacq. on St.John, US Virgin Islands, Lesser Antilles. New Phytol 117:473–481

Benzing DH (1982) Mycorrhizal infections of epiphytic orchids in southern Florida. Am Orchid Soc Bull 51:618–622

Benzing DH (1989a) The evolution of epiphytism. In: Lüttge U (ed) Vascular plants as Epiphytes. Evolution and Ecophysiology. Ecological Studies, vol. 76. Springer, Berlin Heidelberg New York, pp 15–41

Benzing DH (1989b) The mineral nutrition of epiphytes. In: Lüttge U (ed) Vascular plants as epiphytes. Evolution and ecophysiology. Ecological studies, vol 76. Springer, Berlin Heidelberg New York, pp 167–199

Benzing DH (1990) Vascular epiphytes. Cambridge University Press, Cambridge

Benzing DH, Atwood JT (1984) Orchidaceae: ancestral habitats and current status in forest canopies. Syst Bot 9:155–165

Bertsch A (1966) CO_2-Gaswechsel und Wasserhaushalt der aerophilen Grünalge *Apatococcus lobatus*. Planta 70:46–72

Borland AM, Griffiths H, Maxwell C, Broadmeadow MSJ, Griffiths NM, Barnes JD (1992) On the ecophysiology of the Clusiaceae in Trinidad: expression of CAM in *Clusia minor* L. during the transition from wet to dry season and characterization of three endemic species. New Phytol 122:349–357

Broadmeadow MSJ, Griffiths H, Maxwell C, Borland A (1992) The carbon isotope ratio of plant organic material reflects temporal and spatial variations in CO_2 within tropical forest formations in Trinidad. Oecologia 89:435–441

Bruns-Strenge S, Lange O (1992) Photosynthetische Primärproduktion der Flechte *Cladonia portentosa* an einem Dünenstandort auf der Nordseeinsel Baltrum. III. Anwendung des Photosynthesemodells zur Simulation von Tagesläufen des CO_2-Gaswechsels und zur Abschätzung der Jahresproduktion. Flora 186:127–140

Cochard H, Ewers FW, Tyree MT (1994) Water relations of a tropical vine-like bamboo (*Rhipidocladum racemiflorum*) root pressures, vulnerability to cavitation and seasonal changes in embolism. J Exp Bot 45:1085–1089

Coley PD, Kursar TA, Machado J-L (1993) Colonisation of tropical rainforest leaves by epiphylls: effect of site and host plant leaf lifetime. Ecology 74:619–623

Davidson DW, Epstein WW (1989) Epiphytic associations with ants. In: Lüttge U (ed) Vascular plants as epiphytes. Evolution and ecophysiology. Ecological studies, vol. 76. Springer, Berlin Heidelberg New York, pp 200–233

Ehleringer JR, Ullmann I, Lange OL, Farquhar GD, Cowan IR, Schulze ED, Ziegler H (1986) Mistletoes: a hypothesis concerning morphological and chemical avoidance of herbivory. Oecologia 70:234–237

Ewers FW, Fisher JB, Chiu S-T (1990) A survey of vessel dimensions in stems of tropical lianas and other growth forms. Oecologia 84:544–552

Farquhar GD, Ehleringer JR, Hubick KT (1989a) Carbon isotope discrimination and photosynthesis. Annu Rev Plant Physiol Plant Mol Biol 40:503–537

Farquhar GD, Hubick KT, Condon AG, Richards RA (1989b) Carbon isotope fractionation and plant water-use efficiency. In: Rundel PW, Ehleringer JR, Nagy KA (eds) Stable isotopes in ecological research. Ecological studies, vol 68. Springer, Berlin Heidelberg New York, pp 21–40

Franco AC, Ball E, Lüttge U (1992) Differential effects of drought and light levels on accumulation of citric and malic acids during CAM in *Clusia*. Plant Cell Environ 15:821–829

Freiberg ER (1994) Stickstoffixierung in der Phyllosphäre tropischer Regenwaldpflanzen in Costa Rica. Dissertation, Ulm

Galloway DJ (ed) (1991) Tropical lichens: their systematics, conservation, and ecology. The Systematics Association Spec. Vol. No 43. Clarendon Press, Oxford, pp 275–277

Gessner F (1956) Der Wasserhaushalt der Epiphyten und Lianen. In: Ruhland W (ed) Handbuch der Pflanzenphysiologie, Bd III. Pflanze und Wasser. Springer, Berlin Göttingen Heidelberg, pp 915–950

Givnish TJ, Burkhardt EL, Happel RE, Weintraub JD (1984) Carnivory in the bromeliad *Brocchinia reducta*, with a cost/benefit model for the general restriction of carnivorous plants to sunny, moist, nutrient-poor habitats. Am Nat 124:479–497

Goh CJ, Kluge M (1989) Gas exchange and water relations in epiphytic orchids. In: Lüttge U (ed) Vascular plants as epiphytes. Evolution and ecophysiology. Ecological studies, vol 76. Springer, Berlin Heidelberg New York, pp 137–166

Green TGA, Lange OL (1991) Ecophysiological adaptations of the lichen genera *Pseudocyphellaria* and *Sticta* to south temperate rainforests. Lichenologist 23:267–282

Green TGA, Lange OL (1994) Photosynthesis in poikilohydric plants: A comparison of lichens and bryophytes. In: Schulze E-D, Caldwell MC (eds) Ecophysiology of photosynthesis. Ecological studies, vol 100. Springer, Berlin Heidelberg New York, pp 319–341

Green TGA, Kilian E, Lange O (1991) *Pseudocyphellaria dissimilis*:a desiccation-sensitive, highly shade-adapted lichen from New Zealand. Oecologia 85:498–503

Griffiths H (1989) Carbon dioxide concentrating mechanisms and the evolution of CAM in vascular epiphytes. In: Lüttge U (ed) Vascular Plants as epiphytes: evolution and ecophysiology. Ecological studies, vol 76. Springer-Verlag, Berlin Heidelberg New York, pp 42–86

Griffiths H, Smith JAC (1983) Photosynthetic pathways in the Bromeliaceae of Trinidad: relations between life forms, habitat preference and the occurrence of CAM. Oecologia 60:176–184

Haag-Kerwer A, Franco AC, Lüttge U (1992) The effect of temperature and light on gas exchange and acid accumulation in the C_3-CAM plant *Clusia minor* L. J Ex Bot 43:345–352

Holbrook NM, Putz FE (1996a) Physiology of tropical vines and hemiepiphytes: plants that climb up and plants that climb down. In: Mulkey SS, Chazdon RL, Smith AP (eds) Tropical forest plant ecophysiology. Chapman and Hall, New York, pp 363–394

Holbrook NM, Putz FE (1996b) From epiphyte to tree:differences in leaf structure and leaf water relations associated with the transition in growth form in eight species of hemiepiphytes. Plant Cell Environ 19:631–642

Holthe PA, Sternberg L da SL, Ting IP (1987) Developmental control of CAM in *Peperomia scandens*. Plant Physiol 84:743–747

Johansson DR (1975) Ecology of epiphytic orchids in West African rain forests. Am Orchid Soc Bull 44:125–136

Johansson DR (1977) Epiphytic orchids as parasites of their host trees. Am Orchid Soc Bull 46:703–707

Kress WJ (1989) The systematic distribution of vascular epiphytes. In: Lüttge U (ed) Vascular plants as epiphytes. Evolution and ecophysiology. Ecological studies, vol 76. Springer, Berlin Heidelberg New York, pp 234–261

Lange OL, Kilian E, Ziegler H (1986) Water vapor uptake and photosynthesis of lichens: performance differences in species with green and blue-green algae as phycobionts. Oecologia 71:104–110

Lange OL, Green TGA, Ziegler H (1988) Water status related photosynthesis and carbon, isotope discrimination in species of the lichen genus *Pseudocyphellaria* with green or blue-green photobionts and in photosymbiodemes. Oecologia 75:494–501

Lange OL, Büdel B, Heber U, Meyer A, Zellner H, Green TGA (1993) Temperate rainforest lichens in New Zealand:high thallus water content can severely limit photosynthetic CO_2 exchange. Oecologia 95:303–313

Lange OL, Büdel B, Zellner H, Zotz G, Meyer A (1994) Field measurements of water relations and CO_2 exchange of the tropical cyanobacterial basidiolichen *Dictyonema glabratum* in a Panamanian rainforest. Bot Acta 107:279–290

Lüttge U (1983) Ecophysiology of carnivorous plants. In: Lange OL, Nobel PS, Osmond CB, Ziegler H (eds) Physiological plant ecology III. Responses to chemical and biological environment. Encyclopedia of plant physiology NS. Springer, Berlin Heidelberg New York, pp 489–517

Lüttge U (1985) Epiphyten: Evolution und Ökophysiologie. Naturwissenschaften 72: 557–566

Lüttge U (1987) Carbon dioxide and water demand:crassulacean acid metabolism (CAM), a versatile ecological adaptation exemplifying the need for integration in ecophysiological work. New Phytol 106:593–629

Lüttge U (1989) Vascular epiphytes:Setting the scene. In: Lüttge U (ed) Vascular plants as epiphytes. Evolution and ecophysiology. Ecological studies, vol 76. Springer, Berlin Heidelberg New York, pp 1–14

Lüttge U (1991) *Clusia*: Morphogenetische, physiologische und biochemische Strategien von Baumwürgern im tropischen Wald. Naturwissenschaften 78:49–58

Lüttge U, Ball E, Kluge M, Ong BL (1986) Photosynthetic light requirements of various tropical vascular epiphytes. Physiol Vég 24:315–331

Lüttge U, Kluge M, Bauer G (1994) Botanik, 2. Aufl. VCH, Weinheim

Mägdefrau K (1956) Paläobiologie der Pflanzen. G Fischer, Jena

Marshall JD, Ehleringer JR (1990) Are xylem-trapping mistletoes partially heterotrophic? Oecologia 84:244–248

Martin CE (1994) Physiological ecology of the Bromeliaceae. Bot Rev 60:1–82

Martius CFP von, 1840–1906: Flora brasiliensis, vol. 1–15. München and Leipzig

Mez, C (1904) Physiologische Bromeliaceen-Studien. I. Die Wasser-Ökonomie der extrem atmosphärischen Tillandsien. Jahrb Wiss Bot 40:157–229

Nadkarni NM (1981) Canopy roots:convergent evolution in rainforest nutrient cycles. Science 124:1023–1024

Nobel PS (1983) Biophysical plant physiology and ecology. Freeman, San Francisco

Orozco A, Rada F, Azocar A, Goldstein G (1990) How does a mistletoe affect the water, nitrogen and carbon balance of two mangrove ecosystem species? Plant Cell Environ 13:941–947

Pate JS, True KC, Kuo J (1991a) Partitioning of dry matter and mineral nutrients during a reproduction cycle of the mistletoe *Amyema linophyllum* (Fenzl.) Tieghem parasitizing *Casuarina obesa* Miq. J Exp Bot 42:427–439

Pate JS, True KC, Rasins E (1991b) Xylem transport and storage of amino acids by S.W. Australian mistletoes and their hosts. J Exp Bot 42:441–451

Pittendrigh CS (1948) The bromeliad-*Anopheles*-malaria complex in Trinidad. I. The bromeliad flora. Evolution 2:58–89

Putz FE, Holbrook NM (1986) Notes on the natural history of hemiepiphytes. Selbyana 9:61–69

Rada F, Jaimez R (1992) Comparative ecophysiology and anatomy of terrestrial and epiphytic *Anthurium bredmeyeri* Schott in a tropical andean cloud forest. J Exp Bot 43:723–727

Rey L, Sadik A, Fer A, Renandiu S (1991) Trophic relations of the dwarf mistletoe *Arcenthobium oxycedri* with its host *Juniperus oxycedrus*. J Plant Physiol 138:411–416

Richards PW (1952) The tropical rainforest. An ecological study. Cambridge University Press, London

Richter A, Popp M, Mensen R, Stewart RG, von Willert DJ (1995) Heterotrophic carbon gain of the parasitic angiosperm *Tapinanthus oleifolius*. Aust J Plant Physiol 22:537–544

Ruinen J (1953) Epiphytosis. A second view on epiphytism. Ann Bogor 1:101–157

Ruinen J (1961) The phyllosphere. I. An ecologically neglected milieu. Plant Soil 15:81–109

Ruinen J (1965) The phyllosphere. III. Nitrogen fixation in the phyllosphere. Plant Soil 22:375–395

Ruinen J (1974) Nitrogen fixation in the phyllosphere. In: Quispel A (ed) The biology of nitrogen fixation. North Holland Publishing, Amsterdam, pp 121–167

Sallé G, Frochot H, Audary C (1993) Le gui. Recherche 24:1334–1342

Schimper AFW (1888) Botanische Mitteilungen aus den Tropen. II. Epiphytische Vegetation Amerikas. G Fischer, Jena

Schmitt AK, Lee HSJ, Lüttge U (1988) The response of the C_3-CAM tree *Clusia rosea*, to light and water stress. I. Gas exchange characteristics. J Exp Bot 39:1581–1590

Schmitt AK, Martin CE, Lüttge U (1989) Gas exchange and water vapour uptake in the atmospheric CAM bromeliad *Tillandsia recurvata* L.:The influence of trichomes. Bot Acta 102:80–84

Schmucker T, Linnemann G (1959) Carnivorie. In: Handbuch der Pflanzenphysiologie, vol XI. Springer, Berlin Göttingen Heidelberg

Schulze E-D, Turner NC, Glatzel G (1984) Carbon, water and nutrient relations of two mistletoes and their hosts: A hypothesis. Plant Cell Environ 7:293–299

Seifriz W (1924) The altitudinal distribution of lichens and mosses on Mt. Gedeh, Java. J Ecol 12:307–313

Sipes DL, Ting IP (1985) Crassulacean acid metabolism and crassulacean acid metabolism modification in *Peperomia camptotricha*. Plant Physiol 77:59–63

Sipman HJM (1989) Lichen zonation in the Parque los Nevados transect. Stud Trop Andean Ecosyst 3:461–483

Sitte P (1991) Morphologie. In: Sitte P, Ziegler H, Ehrendorfer F, Bresinsky A (eds) Strasburger Lehrbuch der Botanik. G Fischer, Stuttgart, pp 13–238

Smith JAC (1989) Epiphytic bromeliads. In: Lüttge U (ed) Vascular plants as epiphytes. Evolution and ecophysiology. Ecological studies, vol. 76: Springer, Berlin Heidelberg New York, pp 109–138

Smith JAC, Griffiths H, Lüttge U (1986a) Comparative ecophysiology of CAM and C_3 bromeliads. I. The ecology of the Bromeliaceae in Trinidad. Plant Cell Environ 9:359–376

Smith JAC, Griffiths H, Lüttge U, Crook CE, Griffiths NM, Stimmel K-H (1986b) Comparative ecophysiology of CAM and C_3 bromeliads. IV. Plant water relations. Plant Cell Environ 9:395–410

Stewart GR, Schmidt S, Handley LL, Turnbull MH, Erskine PD, Joly CA (1995) ^{15}N natural abundance of vascular rainforest epiphytes:implications for nitrogen source and acquisition. Plant. Cell Environ 18:85–90

Tietze M (1906) Physiologische Bromeliaceen-Studien. II. Die Entwicklung der wasseraufnehmenden Bromeliaceen-Trichome. Zeitschrift für Naturwissenschaften, Halle 78:1–50

Ting IP, Bates L, O'Reilly Sternberg L, DeNiro MJ (1985) Physiological and isotopic aspects of photosynthesis in *Peperomia*. Plant Physiol 78:246–249

Treseder KK, Davidson DW, Ehleringer JR (1995) Absorption of ant-provided carbon dioxide and nitrogen by a tropical epiphyte. Nature 375:137–139

Vareschi V (1980) Vegetationsökologie der Tropen. Ulmer, Stuttgart

Walter H, Breckle S-W (1984) Ökologie der Erde, vol. 2. Spezielle Ökologie der tropischen und subtropischen Zonen. G Fischer, Stuttgart

Winter K, Wallace BJ, Stocker GC, Roksandic Z (1983) Crassulacean acid metabolism in Australian vascular epiphytes and some related species. Oecologia 57:129–141

Zotz G, Winter K (1994a) Photosynthesis and carbon gain of the lichen, *Leptogium azureum*, in a lowland tropical forest. Flora 189:179–186

Zotz G, Winter K (1994b) Annual carbon balance and nitrogen-use efficiency in tropical C_3 and CAM epiphytes. New Phytol 126:481–492

Zotz G, Tyree MT, Cochard H (1994) Hydraulic architecture, water relations and vulnerability to cavitation of *Clusia uvitana* Pittier: a C_3-CAM tropical hemiepiphyte. New Phytol 127:287–295

CHAPTER 5

Mangroves

5.1
Contrasts in Salinity

Mangroves are a characteristic and important type of tropical and subtropical forests, with a unique capacity to tolerate large short-term changes of salinity. The name comes from the Spanish "**mangle**" for *Rhizophora,* a mangrove genus, and the English "**grove**". Mangroves may also be considered as "**tide-forests**", since their ecology is determined primarily by the tides at the 3 typical sites where they occur (Fig. **5.1**):

- coastal mangroves,
- estuarine mangroves and
- coral mangroves,

i.e. mangroves along the coast lines, in river estuaries and around coral reefs and coral islands. However, salinity in mangroves is not only influenced by the **tides**, but also by the **climate**. At high tide salinity in the rooting medium of mangroves, of course, will be determined by sea water. At low tide, however, it will be higher or lower depending on the climatic conditions, i.e.

- **humid climate with rainfall frequently diluting and leaching salt;**
- **arid climate with salt normally concentrated.**

Trees in mangrove forests may become quite tall although often mangroves have a scrub-like physiognomy (e.g. compare Fig. **5.1B** and **F**). Lin and Sternberg (1992a,b, 1993) have compared scrub and tree life-forms of the mangrove species *Rhizophora mangle*. Trees are formed in the fringe forest at lower levels (24 cm above sea level) and the scrub formation at higher levels (60 cm a.s.l.). The scrub form is associated with high salinities occurring at the higher levels during the dry season. In the rainy season, the scrub mangroves can also take up fresh water from rain, and *R. mangle* is therefore a facultative halophyte. However, frequent stress is caused by the changes in salinity following shifts between flooding by ocean water and fresh water introduced by rain. Such variations can lead to a significant decrease in photosynthesis and plant growth in the scrub mangroves, in contrast to constant salinity which maintains the salt load in the substratum.

Fig. 5.1A-F. Coastal mangroves of the Morocoy National Park near Tucacas at the Caribbean coast of Venezuela (**A**, **B**) and in Queensland, Australia (**C**). Estuary mangroves in Trinidad (**D**) and Costa Rica (**E** Las Baulas, **F** Rio Parrita)

Fig. 5.1D-F

Fig. 5.2.
The mangrove fern
Acrostichum aureum
(Costa Rica)

A striking feature of mangrove sites often is the vigorous growth of large terrestrial ferns of the genus *Acrostichum* (Fig. **5.2**). These "mangrove ferns" are shade-tolerant plants, which, however, have their maximum development and productivity under full exposure. *Acrostichum aureum* is quite salt tolerant, although perhaps somewhat less than the mangrove trees, and in particular the gametophytes need reduced salinity during establishment (Medina et al. 1990).

Mangroves delimit most tropical coast-lines and also extend into the sub-tropics (Fig. **5.3**), such that 60-75 % of all tropical coast-lines are occupied by mangroves. The area covered by mangroves is 140 000 km², which is about 0.1 % of the total land surface of the earth. There are about 60 different species of mangroves in several families, some of which are depicted in Fig. **5.4** and listed in Table **5.2** (see below). Floristic diversity is much larger in Asia and Australia than in Africa and America (Fig. **5.3**), although in general terms mangroves are floristically much poorer than other tropical forests.

Although some mangrove trees provide useful and particularly **resistant wood**, mangroves have been frequently considered to be useless and are dis-

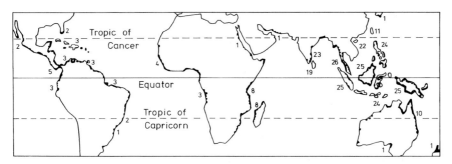

Fig. 5.3. Global distribution of mangroves with *numbers* indicating the approximate number of tree and shrub species in the mangrove vegetation. (After Vareschi 1980, with kind permission of R Ulmer)

Fig. 5.4. Co-occurring mangrove species in the Morocoy Park of Venezuela (see Fig. 5.1A,B): *Avicennia germinans, Laguncularia racemosa, Rhizophora mangle*

appearing rapidly. Only recently it has been realized, that they are ecosystems of not only unique and characteristic beauty but also of great importance for stabilizing coastlines and for **breeding of marine life**, which provides the economic basis of coastal fisheries. Mangroves are among the most productive ecosystems of the world. **Root associations** of mangroves **with halotolerant N_2-fixing bacteria** have been also shown to improve N-supply and contribute to the high productivity (Zuberer and Silver 1979; Sengupta and Chaudhuri 1991).

Some mangrove species are **viviparious** (Fig. 5.5). After fertilization they develop from the zygotes as embryos and then seedlings, which grow out of the flowers and fruits and remain for some while on the mother plant. Once liberated the viviparious seedlings can establish directly in the sediment at low tide or float in the sea water and are dispersed. The advantages of vivipary are not clear since it is observed in only some mangrove families.

Fig. 5.5.
Vivipary in *Rhizophora mangle*

5.2
Inundation of Swampy Soils

In addition to salinity, stress for mangroves results from the inundation of the silty, anaerobic substratum in which they root. The major stress factors thus are

- salinity, and
- low O_2-partial pressure

in the root medium.

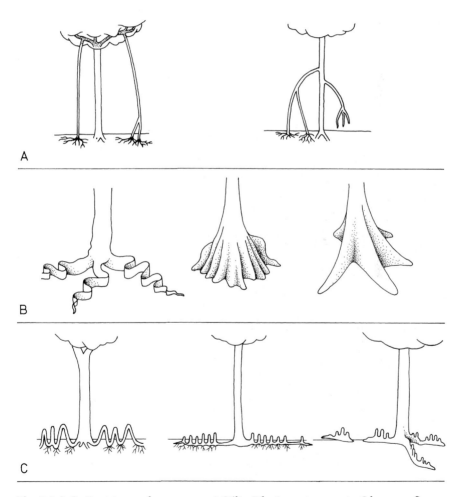

Fig. 5.6 A-C. Root types of mangroves. **A** Stilt-, **B** buttress-type roots; **C** knee- or finger-like pneumatophores. (Vareschi 1980, with kind permission of R. Ulmer)

Fig. 5.7A–D

The surface muds are zones of net heterotrophy. Light limitation beneath mangrove forests means that photosynthesis by benthic microalgae makes a minor contribution to primary productivity (Alongi 1994). Adult mangrove trees develop a diverse range of strangely shaped **root systems** which provide anchorage and aeration in this substratum and hence, these roots are also named **pneumatophores.** Diffusion of gases is highly limited in the inundated soil. Therefore contact of the root system with the atmosphere or with the sea water, depending on tidal level allows gaseous exchange. Figs. **5.6** and **5.7** show a variety of some of the most frequently observed aerial root systems with stilt roots, buttresses and finger- or knee-like protrusions above ground.

The root aeration provided by these pneumatophores is reinforced by a physiological mechanism. The exposed parts of these roots usually have **lenticels,** which are openings in the bark which can be penetrated by gas but not by water. The influx of water is prevented by surface tension in the intercellular spaces of the lenticels. During high tide, root respiration reduces the O_2-concentration in the intercellular spaces of the root aerenchyma to as little as 10 %, i.e. only half of the atmospheric concentration. This causes a **pressure deficit in the aerenchyma,** because the O_2 consumed in respiration can not be reabsorbed readily from the sea water, while the CO_2 liberated can be dissolved as bicarbonate and released. At low tide, when the roots establish contact with the air again, the pressure deficit effectively leads **air** to being **sucked into the root air spaces.**

5.3
Exclusion, Inclusion and Excretion of Salt

Mangroves, like all other halophytes (which are plants growing in saline habitats) utilize strategies where they function as

- **salt excluders** or
- **salt includers** with intracellular **salt dilution** (succulence) and **compartmentation,**

and in addition operate with

- **salt excretion**

(Popp et al. 1993).

Salt exclusion normally only affords resistance against mild or intermediate salinity stress, mainly for osmotic reasons (see Box **4.1**). In order to main-

◀**Fig. 5.7A-D.** Stilt roots of *Rhizophora mangle* (**A**-**C**) and finger-like pneumatophores of *Avicennia germinans* (**D**)

	Xylem sap	Leaves
Rhizophora mucronata without salt glands	17 mM Cl⁻	450 mM Na⁺ 520 mM Cl⁻
Aegialitis annulata with salt glands	100 mM Cl⁻	380 mM NaCl

Table 5.1.
Salt levels in the xylem sap and in adult leaves of two Australian mangrove species in the field. (Atkinson et al. 1967)

tain **osmotic balance** and keep a water potential gradient from the substratum to the plants, salt excluding plants would have to synthezise alternative organic solutes, which would tie-up a large amount of important resources. Thus, the alternative is **salt inclusion** whereby the salt itself is used as a readily available and "cheap" osmoticum. In species which have special **salt glands** on their leaves, surplus salt may be eliminated by **salt excretion**.

Atkinson et al. (1967) performed an experiment illustrating the effectiveness of these mechanisms by comparing two Australian mangrove species in the field (Table **5.1**). One of the two species, *Rhizophora mucronata*, was a salt excluder as shown by the rather low Cl⁻-levels in the xylem sap. The second one, *Aegialitis annulata*, was a salt includer as indicated by the larger xylem sap Cl⁻ concentration. In contrast to the salt excluder, *A. annulata*, has salt glands and is capable of salt excretion. Irrespective of the large differences in xylem sap salt concentrations adult leaves had similar salt levels in both species. Hence, the different strategies for dealing with salt, whether by exclusion at the root level or excretion at the leaf level and dilution via succulence (see below) lead to the same salt level in leaves.

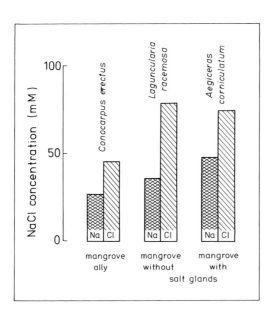

Fig. 5.8.
Na⁺- and Cl⁻ concentrations in the xylem sap of mangroves (*Laguncularia racemosa* without salt glands; *Aegiceras corniculatum* with salt glands) and the mangrove ally *Conocarpus erectus* grown in seawater in the laboratory. (Data from Polanía 1990 and pers. comm. of M. Popp)

However, the distinction between salt excluders and salt includers is only relative. There is always some control of salt uptake at the root level. This is the case in all mangroves, and the **salt concentration in the xylem sap** is always much smaller than in seawater, where the Cl⁻ concentration is over 500 mM (Scholander 1968; Fitzgerald and Allaway 1991). The low xylem sap Cl⁻-content is seen in Table **5.1**, with mangroves in the field, and in Fig. **5.8** with the mangroves *Laguncularia racemosa* and *Aegiceras corniculatum* and the mangrove ally *Conocarpus erectus* grown in seawater in the laboratory. Strictly speaking, therefore, when compared to the sea water rooting

Table 5.2.
Na⁺ and Cl⁻ levels in different mangrove species collected from all over the world. (Popp et al. 1984)

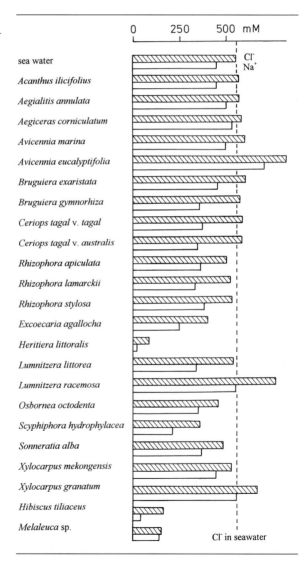

medium, all mangroves are salt excluders. This can add to the salinization of the substratum, which, as mentioned above has ecophysiological implications for photosynthetic CO_2-uptake and transpiration (Passioura et al. 1992; see Sect. **5.4**). The data of Fig. **5.8** also show little difference between NaCl concentrations in the xylem of mangroves with and without salt glands in the leaves. These observations do not support generalisations from the observations of Atkinson et al. (1967) that salt excreting mangroves take up more salt than non-excreting species.

Table **5.2** gives a compilation of NaCl-levels in 23 different mangrove species collected from all over the world. Generally, the **salt concentrations in the leaves** were similar to that of seawater (see also Table **5.1**).

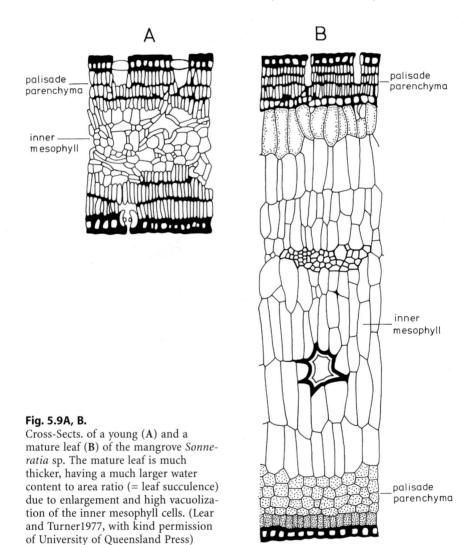

A   **B**

palisade parenchyma

inner mesophyll

palisade parenchyma

inner mesophyll

palisade parenchyma

Fig. 5.9A, B.
Cross-Sects. of a young (**A**) and a mature leaf (**B**) of the mangrove *Sonneratia* sp. The mature leaf is much thicker, having a much larger water content to area ratio (= leaf succulence) due to enlargement and high vacuolization of the inner mesophyll cells. (Lear and Turner1977, with kind permission of University of Queensland Press)

Salt accumulation as a consequence of salt inclusion and the concentrating effect of transpiration has important correlates at the cellular level, namely **salt compartmentation and dilution**. Salt is sequestered (compartmented) in the cell sap vacuoles where it can be diluted by osmotic uptake of water. However, this requires an increased volume if the overall salt concentration were to be maintained at a constant level. Therefore, such salt dilution is associated with succulence ("**salt succulence**"), with the formation of large central vacuoles. This supports maintenance of water relations and turgor pressure according to the relationship

$$\Delta \psi = \Delta P - \Delta \pi, \tag{5.1}$$

where ψ is water potential, P turgor potential and π osmotic potential (see Box **4.1**). Succulence of mangrove leaves may increase as leaves age, and this is mainly due to an enlargement of leaf cells, providing larger vacuoles in which salt can be accumulated and diluted to some extent (Fig. **5.9**). Thus, the total chloride content of leaves, when expressed on a leaf area basis, increases considerably, whereas chloride concentration remains rather constant as succulence increases. This clearly demonstrates the **dilution effect enabled by succulence** (Fig. **5.10**).

As for all halophytes, the cytoplasm, the enzymes and membranes of mangrove cells are as equally sensitive to higher Na^+-concentrations as those of glycophytes. Due to their large hydration shells, Na^+ ions disturb

Fig. 5.10.
Cl⁻-content (mol m⁻² leaf surface) and Cl⁻-concentration (mol l⁻¹ tissue water at saturation) in leaves of the mangrove *Laguncularia racemosa* related to the degree of succulence. The latter is given by the ratio of the leaf-water content at water saturation and the leaf surface (kg m⁻²). (After Biebl and Kinzel 1965, from Kinzel 1982, with kind permission of the author and R. Ulmer)

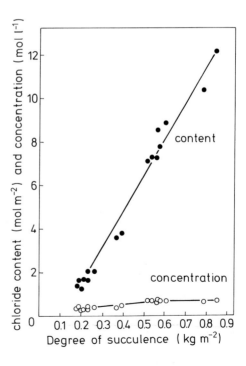

Box 5.1

Compatible Solutes

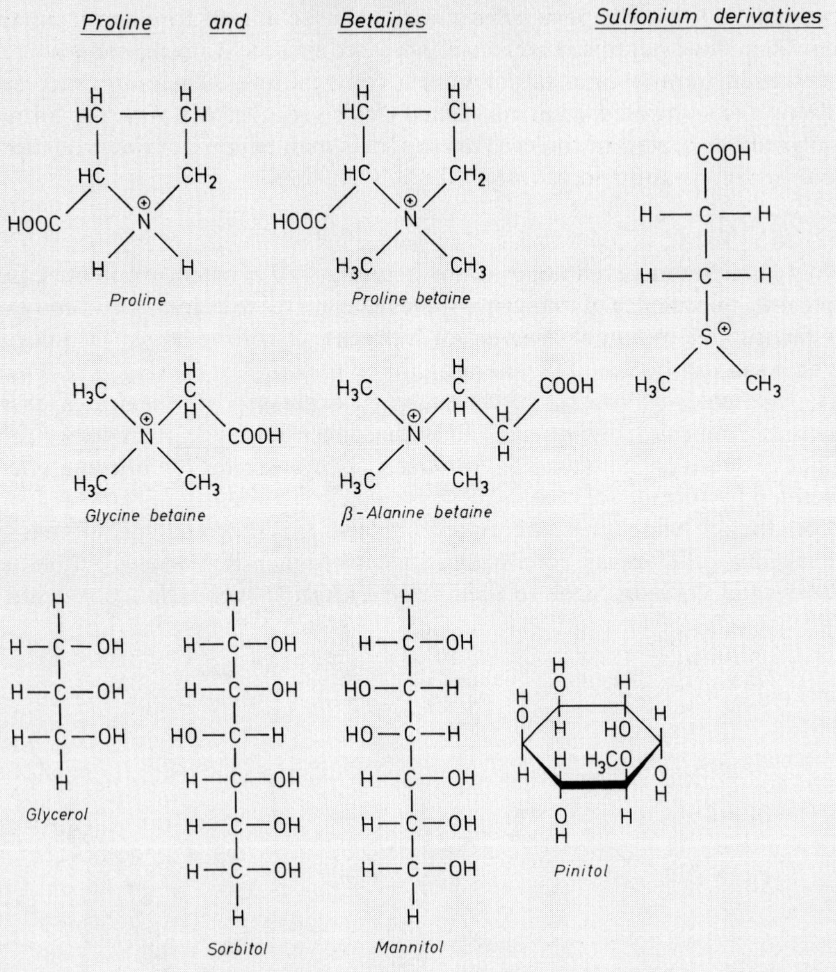

Proline *and* *Betaines* *Sulfonium derivatives*

Proline

Proline betaine

Glycine betaine

β-Alanine betaine

Glycerol

Sorbitol

Mannitol

Pinitol

the molecular water structures, i.e. the specific arrangement of the dipole molecules of H_2O at the surfaces of proteins and membranes. This leads to the requirement for **compartmentation**. The NaCl taken up is sequestered in the vacuoles, where it is accumulated and may be effectively diluted as shown above. However, an osmotic balance is required in the cytoplasm because turgor pressure [Eq. (**5.1**)] can only build up at the plasmalemma/cell wall boundary. The tonoplast membrane itself does not offer enough elastic resistance, and there cannot be a gradient of π across the tonoplast, i.e. $\pi_{cytoplasm}$ must equal $\pi_{vacuole}$. To this end, halophytes normally synthezise small organic molecules which serve as **osmolytes,** and are also called **compatible solutes,** as they both serve as cytoplasmic osmotica and are compatible with water structures. Their function in stabilizing cytoplasmic water structures in fact is very important under salt stress. Box **5.1** presents a variety of compounds, which are known to function as compatible solutes. Sorbitol, mannitol and pinitol are particularly frequent among mangroves (Popp 1984; Popp and Polania 1989; Richter et al. 1990). Accumulation of compatible solutes in the cytoplasm alone is much more efficient in terms of resources and energy needed than if organic molecules were used throughout the whole cell as osmotica to withstand salt stress of the medium. In succulent tissues the relative volume of the cytoplasm is only 1–2 % of the total cell volume, so that vacuolar salt accumulation accompanied by cytoplasmic accumulation of compatible solutes is a very cost effective mechanism of osmotic adjustment.

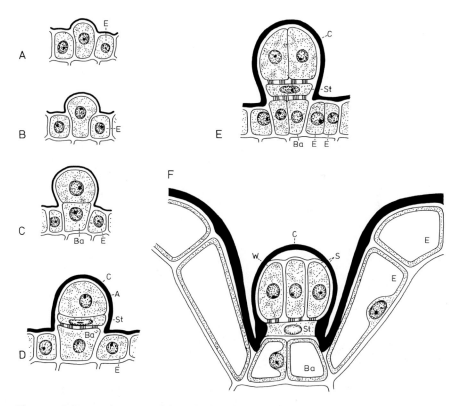

Fig. 5.11A-F. Development of the salt gland hairs of the mangrove *Avicennia marina*. A-E various stages of development, F Mature salt gland (Fahn and Shimony 1977, with kind permission of the author and Linnean Society). A Terminal cell; *Ba* basal cell; *C* cuticle; *E* epidermal cell; *S* secretory cell; *St* stalk cell; *W* cell wall

Fig. 5.12. Leaves of *Avicennia germinans* with salt crystals (*above*) and dissolving salt at high air humidity (*below*)

The mechanism of **salt excretion** by glands has been studied extensively in non-tropical halophytes (Lüttge 1975). It is an energy dependent, active transport process, moving ions against large gradients of their electro-chemical potential. Fig. **5.11** shows the development of the glandular hairs of the mangrove *Avicennia marina* (Fahn and Shimony 1977). The mature **salt gland** is covered and encircled by a cuticle, so that an apoplastic, cell-wall route of salt excretion is not available (Fitzgerald and Allaway 1991). The salt is moved via basal cells, often called "collecting cells", and stalk cells to the secretory cells, which excrete it into a subcuticular space at the head of the gland. Water follows osmotically. The pressure of the excreted fluid increases in the subcuticular space, and the salt is eventually released through pores in the cuticle opening under the hydrostatic pressure. During hot and dry days numerous salt crystals form on the leaves as the excreted salt solution dries (Fig. **5.12**). Conversely, in the early morning, when air humidity is high, the excreted salt on the leaf surface hygroscopically absorbs water and a salty "rain" may drip down from the mangrove trees.

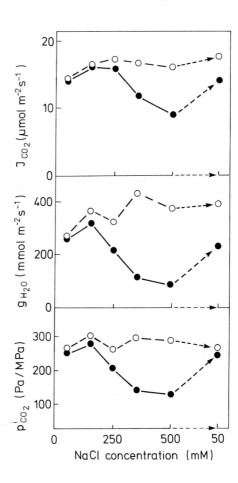

Fig. 5.13.
Net CO_2-uptake (J_{CO_2}), leaf conductance for water vapour (g_{H_2O}) and internal CO_2 partial pressure ($p^i_{CO_2}$) in leaves of *Avicennia marina* at increasing salinity (*closed symbols*) and upon return from high (500 mM NaCl) to low salinity (50 mM NaCl). *Open symbols* controls not treated with NaCl. (After Ball and Farquhar 1984b)

5.4
Gas Exchange

Gas exchange has been studied in relation to the degree of **substratum salinity** by two groups, *viz.* Ball and Farquhar (1984a,b) and Clough and Sim (1989). Fig. **5.13** shows for *Avicennia marina* that net CO_2-uptake (J_{CO_2}), leaf conductance for water vapour (g_{H_2O}) and CO_2-partial pressure in the internal air spaces of the leaves ($p^i_{CO_2}$) are unaffected by salinity up to about 150 mM NaCl and then they decrease at higher degrees of salinity. However, appreciable photosynthesis is still maintained at 500 mM NaCl, where CO_2-uptake rate still is 9 μmol m^{-2} s^{-1}. The responses are reversible upon return from high (500 mM NaCl) to low salinity (50 mM NaCl).

Other measurements with *Avicennia marina* and *Avicennia corniculatum* showed that in addition to substratum salinity **leaf-air water-vapour-pressure difference** (ΔW) also regulates gas exchange. At all salinity levels J_{CO_2}, g_{H_2O} and $p^i_{CO_2}$ also declined with increasing ΔW. **Water-use-efficiencies** (WUE defined as the ratio of CO_2 assimilated to H_2O transpired) decreased with increasing salinity and increasing ΔW (Fig. **5.14**). However, Ball (1986) argues that in comparison to other C_3-plants all WUE-values obtained in Fig. **5.14** are, exceptionally high. Thus, healthy mangrove leaves can operate at comparatively low internal CO_2 concentrations, i.e. at $p^i_{CO_2}$ of about 170 Pa/MPa, while C_3-crop plants for example need to maintain 220 Pa/MPa. This allows mangroves to control water use while maintaining photosynthesis more than in other C_3-plants.

Clough and Sim (1989) have studied several mangrove species at different sites in Australia and Papua New Guinea, which provided a natural gradient of salinity (Fig. **5.15**). This confirmed the correlations between J_{CO_2}, g_{H_2O} and $p^i_{CO_2}$, with salinity as well as with ΔW already seen in the work of Ball and Farquhar (1984a,b). These authors also calculated **intrinsic water use efficiency**, "W", as

$$\text{"W"} = (p^a_{CO_2} - p^i_{CO_2}) / (p^a_{CO_2} - \Gamma), \tag{5.2}$$

where $p^a_{CO_2}$ and $p^i_{CO_2}$ are ambient and internal CO_2-partial pressure, respectively, and Γ is the CO_2-compensation point of photosynthesis. It is evident from equation (5.2), that at constant $p^a_{CO_2}$ and Γ, "W" is inversely reated to $p^i_{CO_2}$, and increases with salinity as shown by the data of Clough and Sim (1989) in Fig. **5.15**. Similar results were found by Ball and Farquhar (1984a,b), where $p^i_{CO_2}$ also declined with salinity as well as with ΔW. When WUE is calculated as the assimilation to transpiration ratio, stomata regulate the resistance to diffusion of CO_2 and water vapour, but in addition the external factor of atmospheric water-vapour pressure-deficit plays an essential part as the driving force for transpirational water loss from the leaves. In the derivation of "W" this external factor is not involved and intrinsic stomatal control is reflected in $p^i_{CO_2}$. While there is a decline in WUE with increasing salinity, in contrast intrinsic water use efficiency, "W", since it is

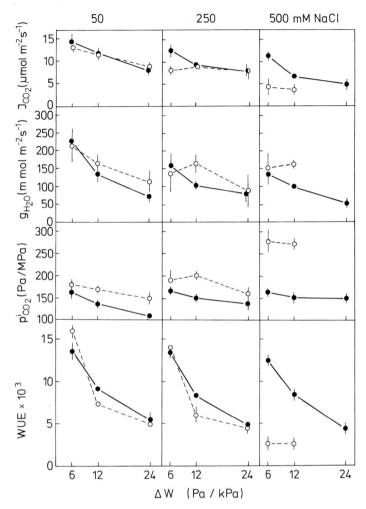

Fig. 5.14. Net CO_2 uptake (J_{CO2}), leaf conductance for water vapour (g_{H_2O}), internal CO_2 partial pressure ($p^i_{CO_2}$) and water-use efficiency (*WUE*) in leaves of *Avicennia marina* (●) and *Avicennia corniculatum* (o) at increasing salinity and leaf-air water-vapour pressure difference (ΔW). (After Ball and Farquhar 1984a)

based on the CO_2-compensation point and $p^i_{CO_2}$, may even suggest improved use of water as stomata close and salinity and ΔW increase.

The relationships discussed above may be interpreted as illustrating the compromise of the **desiccation-starvation dilemma** (see Sect. 3.7.3). Since the flow of salt into the leaves is proportional to salinity and transpiration, control of transpiration also reduces the salt load and the danger of serious water deficits and salt toxicity in the leaves. The reduction of CO_2-gain, which also follows partial stomatal closure, is partially offset by effective

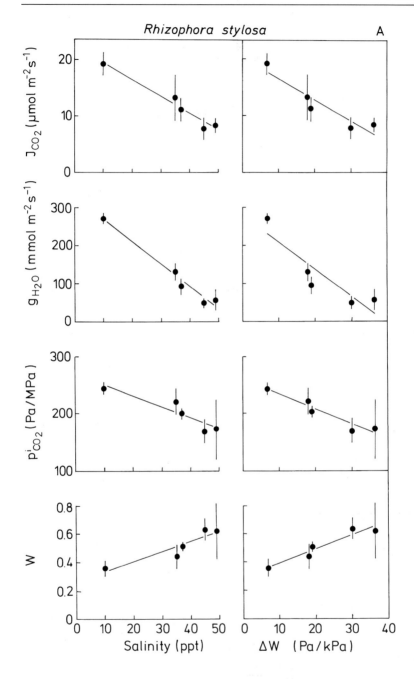

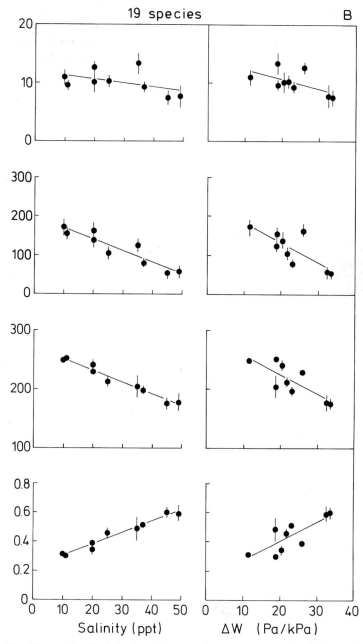

Fig. 5.15A, B. Net CO_2 uptake (J_{CO_2}), leaf conductance for water vapour (g_{H_2O}), internal CO_2 partial pressure ($p^i_{CO_2}$) and intrinsic water-use-efficiency (W) along natural gradients of salinity and leaf-air water-vapour-pressure difference (ΔW) for *Rhizophora stylosa* (**A**) and for 19 different mangrove species (**B**) averaged for each locality. Soil salinity is given in ppt, where 30-35 ppt correspond to the salinity of seawater (0.52-0.55 M NaCl). (Clough and Sim 1989)

CO_2-fixation at low $p^i_{CO_2}$, whilst maintaining high WUE (as noted by Ball 1986; see above). Interestingly, it was also observed that of all the mangroves studied, the salt-excreting species (*Avicennia marina*) afford the highest rates of CO_2-uptake and water-vapour conductances, although the salt load of leaves may be similar in excreting and non-excreting species (Table **5.1**).

Fig. 5.16A, B. *Avicennia germinans* (**A**) and *Conocarpus erectus* (**B**) on an alluvial sand plain at the Caribbean coast of Venezuela

Comparative studies of *Conocarpus erectus* and *Avicennia germinans* during the rainy season and the dry season offer additional insights into the success of mangroves under strongly varying conditions of salinity. *C. erectus* is not a true mangrove, but rather a mangrove associate or ally. It does not grow as close to salt-water lagoons and estuaries or tidal plains as the mangroves *sensu strictu*. However, at places its distribution overlaps with that of mangroves, for instance in alluvial sand plains at the Caribbean coast of Venezuela. There are various sizes of vegetation islands on these sand plains (see Chap. **6**), where *C. erectus* and *A. marina* are found in close proximity (Fig. **5.16**). In the rainy season the sand plains may be flooded by fresh water to a depth of half a meter, but in the dry season they dry out becoming hypersaline and covered with a crust of salt (Chap. **6.1**).

These two species were found here growing on the same vegetation island, and Fig. **5.17** shows that in the rainy season, both species had similar CO_2-uptake rates, J_{CO_2}, but that *A. germinans* operated at considerably lower conductance, g_{H_2O}, and internal CO_2-concentration, $p^i_{CO_2}$, than *C. erectus*. In the dry season, J_{CO_2} in the morning was similar to that measured in the wet season for *A. germinans*. There was a midday depression (see Sect. **3.7.2**), which was followed, however, by considerable recovery in the afternoon; $p^i_{CO_2}$ was similar to that in the wet season although g_{H_2O} was somewhat reduced. Conversely, CO_2-uptake in *C. erectus* in the dry season was greatly reduced, with only a small peak in the morning, and g_{H_2O} and $p^i_{CO_2}$ were low throughout the day. It is evident that with similar J_{CO_2} for both species in the rainy season, the smaller reduction of J_{CO_2} in the dry season allowed *A. germinans* to maintain productivity under salinity-stress and drought better than the mangrove-associate *C. erectus*. This illustrates the capacity of the true mangrove *A. germinans* to operate at low $p^i_{CO_2}$ (see above and Ball 1986) as compared to the mangrove ally *C. erectus*. Thus, the true mangrove maintained similar rates of J_{CO_2} to *C. erectus* at lower $p^i_{CO_2}$ in the rainy season and then J_{CO_2} in *C. erectus* was greatly reduced as $p^i_{CO_2}$ declined in the dry season (Fig. **5.17**).

C. erectus does not grow close to the shoreline or reach the tidally inundated mud plains, but it may grow around sand dunes. It is then subject to salinity from salt spray and develops very thick succulent leaves at the windward side of the bushes, while leaves on the sheltered side are non-succulent (Fig. **5.18**). The non-succulent leaves protected from the salt spray have higher J_{CO_2} and transpiration, J_{H_2O}, during the second part of the day than the salt-exposed leaves (Fig. **5.19**).

Due to their growth on saline substrata and in exposure (see Fig. **5.1**), with generally low leaf-conductance for water vapour and CO_2-uptake rates and high water-use-efficiencies, mangroves are subject to **photoinhibition** and need **photoprotection**. Lovelock and Clough (1992) have shown that among mangroves of the Daintree River in Australia (17° S, 147° E) there are different strategies. In *Rhizophora* there is stress avoidance as leaves are oriented nearly vertically and thus reduce light absorption. *Bruguiera par-*

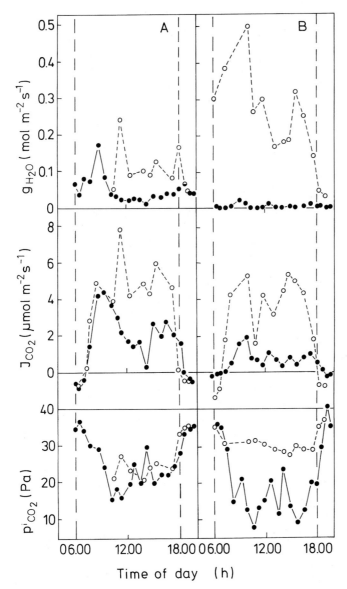

Fig. 5.17A, B. Leaf conductance for water vapour, g_{H_2O}, net CO_2-uptake, J_{CO_2}, and internal CO_2 partial pressure, $p^i_{CO_2}$, in leaves of the same plants of *Avicennia germinans* (**A**) and *Conocarpus erectus* (**B**) studied during the rainy season (o) and the dry season (●). (After Smith et al. 1989)

Fig. 5.18A, B. Wind shaped bushes of *Conocarpus erectus* on sand dunes of the Paraguana Peninsula (**A**) and Chichiriviche (**B**) on the Caribbean coast of Venezuela.

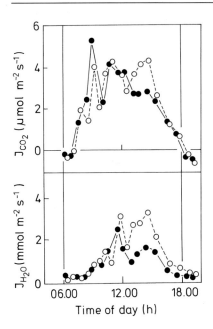

Fig. 5.19.
Net CO_2-uptake, J_{CO_2}, and transpiration, J_{H_2O}, of *Conocarpus erectus* on coastal sand dunes. ● Succulent leaves on the windward side of the bushes, o sheltered, on the leeward side. (Smith et al. 1989)

viflora has small horizontal leaves, which are rich in xanthophylls functioning in dissipation of excitation energy (Sect. **3.6.2.4**). Larger horizontally arranged leaves of *Bruguiera gymnorhiza* tend to heat up more strongly and are therefore more subject ot photodamage (Sect. **3.6.2.4**). Thus, *B. parviflora* dominates the canopy, whereas *B. gymnorhiza* is less abundant at the top of the canopy.

References

Alongi DM (1994) Zonation and seasonality of benthic primary production and community respiration in tropical mangrove forests. Oecologia 98:320–327

Atkinson MR, Findlay GP, Hope AB, Pitman MG, Saddler HDW, West KR (1967) Salt regulation in the mangroves *Rhizophora mucronata* Lam. and *Aegialitis annulata* R.Br. Aust J Biol Sci 20:589–599

Ball MC (1986) Photosynthesis in mangroves. Wetlands (Aust) 6:12–22

Ball MC, Farquhar GD (1984a) Photosynthetic and stomatal responses of two mangrove species, *Aegiceras corniculatum* and *Avicennia marina*, to long term salinity and humidity conditions. Plant Physiol 74:1–6

Ball MC, Farquhar GD (1984b) Photosynthetic and stomatal responses of the grey mangrove, *Avicennia marina*, to transient salinity conditions. Plant Physiol 74:7–11

Biebl R, Kinzel H (1965) Blattbau und Salzgehalt von *Laguncularia racemosa* (L) Gaertn. f. und anderer Mangrovebäume auf Puerto Rico. Österr Bot Z 112:56–93

Clough BF, Sim RG (1989) Changes in gas exchange characteristics and water use efficiency of mangroves in response to salinity and vapour pressure deficit. Oecologia 79:38–44

Fahn A, Shimony C (1977) Development of the glandular and non-glandular leaf hairs of *Avicennia marina* (Forsskål) Vierh. Bot J Linn Soc 74:34–46

Fitzgerald MA, Allaway WG (1991) Apoplastic and symplastic pathways in the leaf of the grey mangrove *Avicennia marina* (Forssk.) Vierh New Phytol 119:217–226

Kinzel H (1982) Pflanzenökologie und Mineralstoffwechsel. Ulmer, Stuttgart

Lear R, Turner T (1977) Mangroves of Australia. University of Queensland Press, St Lucia

Lin G, Sternberg L da SL (1992a) Effect of growth form, salinity, nutrient and sulfide on photosynthesis, carbon isotope discrimination and growth of red mangrove (*Rhizophora mangle* L.). Aust J Plant Physiol 19:509–517

Lin G, Sternberg L da SL (1992b) Comparative study of water uptake and photosynthetic gas exchange between scrub and fringe red mangroves *Rhizophora mangle* L. Oecologia 90:399–403

Lin G, Sternberg L da SL (1993) Effects of salinity fluctuation on photosynthetic gas exchange and plant growth of the red mangrove (*Rhizophora mangle* L.). J Exp Bot 44:9–16

Lovelock CF, Clough BF (1992) Influence of solar radiation and leaf angle on leaf xanthophyll concentrations in mangroves. Oecologia 91:518–525

Lüttge U (1975) Salt glands. In: Baker DA, Hall JL (eds) Ion transport in plant cells and tissues. North-Holland Publishing, Amsterdam, pp 335–376

Medina E, Cuevas E, Popp M, Lugo AE (1990) Soil salinity, sun exposure, and growth of *Acrostichum aureum*, the mangrove fern. Bot Gaz 151:41–49

Passioura JB, Ball MC, Knight JH (1992) Mangroves may salinize the soil and in so doing limit their transpiration rate. Funct Ecol 6:476–481

Polanía J (1990) Anatomische und physiologische Anpassungen von Mangroven. Dissertation, Vienna

Popp M (1984) Chemical composition of Australian mangroves. II. Low molecular weight carbohydrates. Z Pflanzenphysiol 113, 411–421

Popp M, Polanía J (1989) Compatible solutes in different organs of mangrove trees. Ann Sci For 46:842s–844s

Popp M, Larher F, Weigel P (1984) Chemical comparison of Australian mangroves. III. Free amino acids, total methylated onium compounds and total nitrogen. Z Pflanzenphysiol 114:15–25

Popp M, Polanía J, Weiper M (1993) Physiological adaptations to different salinity levels in mangrove. In: Lieth H, Al Masoom A (eds) Towards the rational use of high salinity tolerant plants, vol 1. Kluwer, Dordrecht, pp 217–224

Richter A, Thonke B, Popp M (1990) 1S-10-methylmucoinositol in *Viscum album* and members of the Rhizophoraceae. Phytochemistry 29:1785–1786

Scholander PF (1968) How mangroves desalinate sea water. Physiol Plant 21:251–261

Sengupta A, Chaudhuri A (1991) Ecology of heterotrophic dinitrogen fixation in the rhizosphere of mangrove plant community at the Ganges River estuary in India. Oecologia 87:560–564

Smith JAC, Popp M, Lüttge U, Cram WJ, Diaz M, Griffiths H, Lee HSJ, Medina E, Schäfer C, Stimmel K-H, Thonke B (1989) Ecophysiology of xerophytic and halophytic vegetation of a coastal alluvial plain in northern Venezuela. VI. Water relations and gas exchange of mangroves. New Phytol 111:293–307

Vareschi V (1980) Vegetationsökologie der Tropen. Ulmer, Stuttgart

Zuberer DA, Silver WS (1979) Nitrogen fixation (acetylene reduction) and the microbial colonization of mangrove roots. New Phyol 82:467–472

CHAPTER 6

Salinas

6.1
Physiognomy of the Ecosystem

Examples of well-known salinas or inland salt marshes are the chotts in North Africa (Fig. **6.1**) or the Salinas Grandes in Argentina. The latter extend over more than 250 km from the SW to the NE and across 75 km. However, these salinas are not really within the tropics, at best they are subtropical. Nevertheless, salinas are also found in the tropics. A typical example are the inland salt-marshes near the northern caribbean coast of Venezuela first described briefly by Walter (Walter and Breckle 1984) and later studied ecophysiologically to some detail (Lüttge et al. 1989a,b; Medina et al. 1989; Smith et al. 1989) as summarized in this chapter.

Such salt marshes are formed in bays, where the sand-laden waves returning from the beach to the sea have been driven back towards the shore by the North-Eastern trade winds. First a sandbar is formed leading to a sandbank and to separation of a lagoon. Subsequently, this fills in with sand, drying out and becoming a salt marsh. Both fixed and mobile sand dunes may also form at the coast behind the salt marsh (Fig. **6.2**).

The most salt-resistant plants found first in such sites are mangroves, which begin to surround the lagoon (Fig. **6.2**), an example being *Avicennia germinans* at the lagoon and salt marshes near Chichiriviche on the caribbean coast of Venezuela. Even the high stress resistance of *A. germinans* was insufficient to protect the plant during an extended drought in the 1930's, with the remnants of the dead mangrove forest still remaining (Fig. **6.3**).

The flat alluvial sand plain covering areas previously occupied by the lagoon is subject to marked seasonality because there is a pronounced rainy season in October – December and a strong dry season during the rest of the year interrupted only by a small and short wet period in April (Fig. **6.4**). During the rainy season the sand plain may be covered by several decimeters of fresh water, whereas during the dry season the surface is dry and a considerable salt crust may form (Fig. **6.5**). The very salty groundwater, with an NaCl-concentration several times that of seawater (Fig. **6.6**), percolates upwards to the surface where the water evaporates and leaves behind the dissolved salt. The vegetation of the sand plain can be described by distinguishing five units (Fig. **6.7**):

Fig. 6.1A, B. Chott Fedjadj (A) and Chott el Djerid (B), Tunisia

Fig. 6.2A-E.
Formation of salt marshes from coastal bays and lagoons. (After Walter and Breckle 1984, with kind permission of S.-W. Breckle and G. Fischer-Verlag)

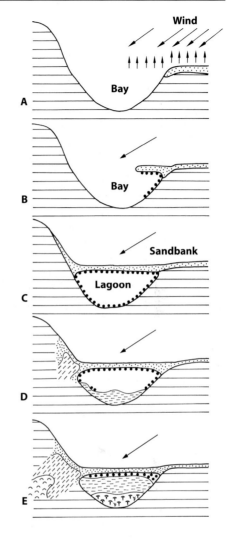

sand

mangroves

salt marsh

fixed sand dunes

mobile sand dunes

thornbush

cacti

a) the vegetation-free sand and salt flats (Fig. **6.7A**),
b) a halophyte zone with *Batis maritima* and *Sesuvium portulacastrum* as the dominating species (Fig. **6.7B**),
c) a grass-land zone with *Sporobulus virginicus* and *Oxycarpha suaedifolia* as the characteristic plants (Fig. **6.7 D**),
d) vegetation islands with the mangrove associate *Conocarpus erectus* and the cactus *Subpilosocereus ottonis* as the physiognomically determinant species (Fig. **6.7 C**),
e) a deciduous forest characterized by species of *Capparis, Caesalpinia coriaria, Prosopis juliflora, Jacquinia revoluta, Maytenus karenii, Erythroxylon cumanense, Croton* sp. and *Pereskia guamacho* (Fig. **6.7D**).

Fig. 6.3. Mangrove forests around the lagoon of Chichiriviche, Venezuela, with remnants of dead trees due to a drought period in 1930s

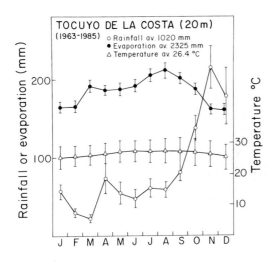

Fig. 6.4.
Average monthly values of rainfall, evaporation and temperature near the northern caribbean coast of Venezuela close to Chichiriviche for a 22-year period. (Medina et al. 1989)

Fig. 6.5A, B. Alluvial sand plain with vegetation islands near the northern carribbean coast of Venezuela at Chichiriviche. **A** In the rainy season (November 1985) covered with fresh water. **B** In the dry season (February 1983) covered with a thick salt crust (background). (See Medina et al. 1989)

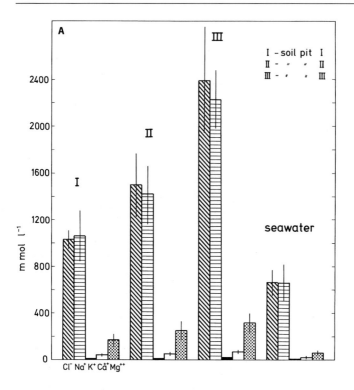

Fig. 6.6A, B.
Ionic composition of the ground water from pits excavated in the dry season on the sand plain at some distance from (*I* and *II*) and close to the open lagoon (*III*) respectively, of Chichiriviche, Venezuela, as compared to seawater. **B** Example of a soil pit dug to examine the ground water.
(Medina et al. 1989)
(For the position of the three soil pits see also the transect of Fig. **6.8**)

Fig. 6.7A-D. Vegetation units of the alluvial plain at Chichiriviche, Venezuela. **A** Vegetation island on the sand plain. **B** Halophyte zone with *Sesuvium portulacastrum* (*front*) and *Batis maritima* (*middle ground*) around a vegetation island with *Conocarpus erectus* in the background. **C** Vegetation island with bushes of *Conocarpus erectus* and *Subpilosocereus ottonis*. **D** Grassland with deciduous forest in the background

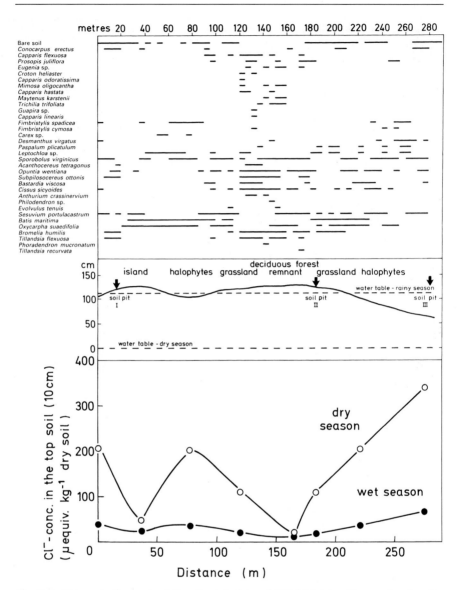

Fig. 6.8. Transect of a part of the alluvial plain of Chichiriviche, Venezuela, showing topographical variations with the major vegetation units (*centre*), the water tables (*centre*) and top soil chloride concentrations in the rainy season and the dry season respectively (*bottom*) and the distribution of the most frequent plant species (*top*). (After Medina et al. 1989)

A transect presenting finer details is shown in Fig. **6.8**. In addition to the seasonal differences in the water table the figure also gives an indication of the seasonal changes in salt content of the upper 10 cm of top-soil. Overall salt concentrations in the soil as well as seasonal fluctuations tend to be highest in the bare areas and decrease with the sequence of vegetation units as follows: halophyte zone – grassland zone – vegetation islands – deciduous forest.

6.2
Dynamics of Vegetation Islands

The most conspicuous feature of these salinas are the small vegetation islands with a diameter of 3 – 10 m and a soil surface 10 – 40 cm higher than the sand plain. Observers have been tempted to consider these islands as a particular stage in a progressive succession, which starts from the bare sand plain, then leads to the halophyte vegetation, followed by island vegetation and finally on to grassland and deciduous forest (Walter 1973).

However, a closer examination extending over several years has shown, that there is no such one-way progressive succession towards a stable climax community at the end. The whole ecosystem is highly dynamic and provides an excellent example of an oscillating mosaic, as opposed to a stable climax equilibrium (see Sect. **3.3.1.3**). By following the development of a given island, we see that islands not only grow into savannas and forest but also die, being eroded and eventually disappearing into the bare sand plain (Fig. **6.9**). Thus, these islands appear to be metastable states between the more stable states of the forest and the sand plain, respectively. Oscillations between the various vegetation units in time may be determined by medium-term climatic fluctuations; e.g. the years between 1966 and 1975 appeared to be wetter, and the years between 1976 and 1986 were drier than the long term average (Fig. **6.10**).

A similar mosaic of bare sand, vegetation islands, shrubs and forests is found in the coastal **restinga** formation of Brazil on sand dunes and on the sand plains behind the shoreline sand dunes. This vegetation is characterized by bromeliads, cacti, orchids and species of *Clusia* with many epiphytes including lichens and many CAM plants (Henriques et al. 1986). However, these sites do not form salt crusts and are no salinas.

Fig. 6.9A–C

Fig. 6.9 A–E A Primordial or decaying vegetation island? **B-E** Continuous observation of a decaying and eroding vegetation island, i.e. the same island in November 1985 (**B**), March 1986 (**C**), April 1987 (**D**) and January 1989 (**E**). North Coast of Venezuela near Chichiriviche

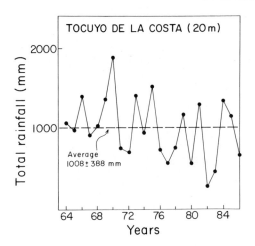

Fig. 6.10
Variations of total annual rainfall near the northern Caribbean coast of Venezuela close to Chichiriviche for the years 1964-1986. (Medina et al. 1989)

6.3
Strategies of Adaptation of Plants in the Different Vegetation Units

6.3.1
Small Perennial Halophytes: Salt Inclusion and Stress Tolerance

The zone of small perennial halophytes surrounding the edges of vegetation islands and bordering the vegetation free salt flats is dominated by three species, namely *Portulaca rubricaulis*, *Sesuvium portulacastrum* and *Batis maritima* (Fig. **6.11**). The creeping fruticose stems of the first two species contrast with *B. maritima*, which has a more upright growth habit. The three species are also primary colonizers of the sand plain. They occupy the most extreme habitats in the sand plain, which range from being flooded with fresh water in the rainy season to dry, salt-encrusted soil in the dry season when the water table may drop to 1 m below ground (Lüttge et al. 1989b; Figs. **6.5** and **6.8**).

All of the three halophytes are **salt includers** and accumulate NaCl in their highly **succulent leaves**. However, *P. rubricaulis* combines this strategy of stress tolerance with that of stress avoidance in that it is **deciduous** and sheds its leaves during the dry season. Among the three species, it has the lowest salt concentrations in its leaf sap, viz. 230 mM Cl^- and 60 mM Na^+, while the other two species have much higher salt levels in their leaves, i.e. 260–1080 mM Cl^- and 370–720 mM Na in the wet season, and 540–1410 mM Cl^- and 920–1590 mM Na^+ in the dry season, respectively. However, *P. rubricaulis* is a **C_4-plant** while the other two species are C_3-plants. Thus *P. rubricaulis* may use the higher instantaneous productivity of C_4-photosynthesis (see box **7.3**) for production of enough perennial shoots to compete in the habitat effectively.

Fig. 6.11A-C
(Continued on page 238)

Fig. 6.11A-E. *Portulaca rubricaulis* (**A** rainy season Nov. 1985); *Sesuvium portula-castrum* (**B** rainy season, Nov. 1985; **C** dry season March 1986); *Batis maritima* (**D** rainy season Nov. 1985; **E** dry season March 1986) in the sand plain of Chichiriviche, Venezuela

B. maritima also accumulated sulphate, with a two-fold increase of leaf-sap concentrations in the dry season. In *S. portulacastrum*, Na^+ accumulation exceeded Cl^- accumulation by far and synthesis of the organic acid anion oxalate is found to serve in maintaining charge balance. Increased salt accumulation in the leaves of *B. maritima* and *S. portulacastrum* in the dry season is accompanied by a 1.5- to 2-fold increase in leaf succulence (see also Chap. **5.3**). *S. portulacastrum* was also shown to use **compatible solutes** (see Chap. **5.3** and Box **5.1**) such as proline and pinitol which augments the tolerance of salt inclusion (Fig. **6.12**).

Fig. 6.12.
Concentrations of the compatible solutes proline and pinitol in the leaf sap of *Sesuvium
portulacastrum* in the wet season (*W*) and the
dry season (*D*). (Lüttge et al. 1989b)

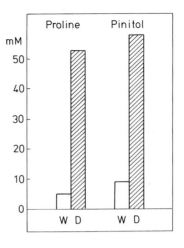

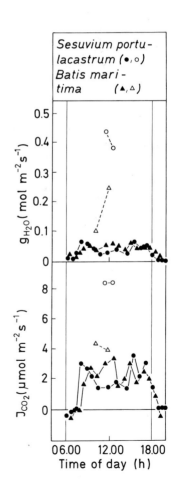

Fig. 6.13
Net CO_2 exchange (J_{CO_2}) and water-vapour
conductance of leaves (g_{H_2O}) of *Sesuvium
portulacastrum* (●, o) and *Batis maritima*
(▲, Δ) in the sand plain of Chichiriviche,
Venezuela, in the wet season (o, Δ) and the
dry season (●, ▲). (Lüttge et al. 1989b)

In leaves of the erect stems of *B. maritima* photosynthetic **gas exchange**, measured as net CO_2 uptake and transpirational loss of water vapour, shows little response to the transitions between the rainy and the dry season (Fig. **6.13**). In contrast, photosynthesis in the leaves of the prostrate stems of *S. portulacastrum* is severely impaired in the dry season, showing a pronounced **midday depression** of gas-exchange (see Sect. **3.7.2**) and about 40 % inhibition of light saturated rates of photosynthesis (Fig. **6.13**). Thus, *S. portulacastrum* clearly suffers more under the stress of the dry season but it is also more of a pioneer coloniser of the sand plain as it occupies the outermost edges of the vegetation islands and larger areas of the flat plain (Fig. **6.7B** and **6.11C**).

In summary, the three species of halophytes, which are subject to very similar challenges by extreme environmental conditions, have different strategies of adaptation to stress. Notwithstanding the similarities in lifeforms, with small fruticose stems and succulent leaves, the differences in ecophysiological comportment reflect ecological diversity, and once again prove to be a basis for species diversity (see Sects. **3.3.1.2** and **3.8**).

6.3.2
Terrestrial CAM Plants: Salt Exclusion and Stress Avoidance

There are two different life forms of terrestrial CAM plants on the sand plain, which are stem succulent cacti and tank forming bromeliads. They use the flexibility of the CAM cycle (see Sect. **3.7.3** and box **3.11**) in strategies of salt exclusion and stress avoidance in the salinas.

6.3.2.1
Columnar Ceroid Cacti

The dominating cactus of the salinas at the North coast of Venezuela is the columnar cactus *Subpilosocereus ottonis* with strongly branching individuals more than 6 m tall (Fig. **6.14 A**). Small seedlings are frequently found at the rim of vegetation islands (Fig. **6.14 B**) as well as among other vegetation and even within the tanks of bromeliads. A similar type of cactus occurs in the subtropical salinas, Salinas Grandes in Argentina, namely *Cereus validus*. Experiments with 10 cm tall seedlings of *C. validus* (Fig. **6.14 D**) and analyses of *S. ottonis* in the field show that these **cacti are** strong **salt excluders** (Fig. **6.15**). When subject to salinity the roots rapidly accumulate large amounts of NaCl. However, there is no salt export from the roots to the shoots, so that the peripheral green stem chlorenchyma and the central water storage parenchyma of the stems receive very little additional salt during a salinity treatment of up to 14 days and 600 mM NaCl in the root medium (Fig. **6.15**).

It is quite obvious that the fine absorptive roots of the cacti die under the stress of salinity and the rest of the cactus becomes quite isolated from the

Fig. 6.14A-F. Large plants (**A**), small seedling (**B**) and seedling rerooted after injury in the field (**C**) of *Subpilosocereus ottonis* in the alluvial plain of Chichiriviche, Venezuela. Experimental seedlings of *Cereus validus* (**D**), one rotting during an extended salt treatment (**E**), and another one dried out at the bottom and totally insulated from the substratum (**F**)

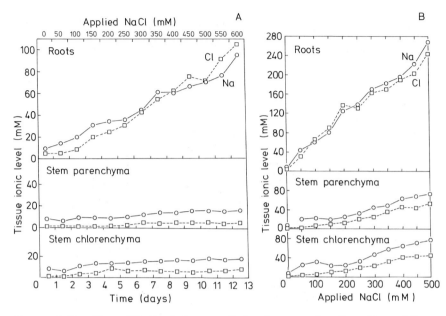

Fig. 6.15A, B. Effects of NaCl solutions supplied to the roots of small seedlings of *Cereus validus* (see Fig. **6.14D-F**) in pot culture with sand on Na⁺ and Cl⁻ levels in the roots, the water storage stem parenchyma and the peripheral green chlorenchyma. **A** NaCl concentrations in the watering solution were increased by daily increments of 50 mM up to 600 mM (*upper abscissa*), and the plants were analyzed as soon as the respective concentrations were reached at the times indicated (*lower abscissa*). **B** NaCl concentrations in the watering solution were increased by daily increments of 50 mM. At any given concentration indicated on the abscissa, plants were kept for 14 days after this concentration was reached and then analyzed. (Nobel et al. 1984)

substratum. When the salinity treatment is extended for several months, in some cases salt solution may diffuse upwards killing the stem tissue so that eventually the whole cactus seedling rots away and dies (Fig. **6.14E**). In other cases, however, the base of the shoot dries out and then finally becomes insulated very effectively from the ground (Fig. **6.14F**). The major part of the green stem survives. But by what means and for how long?

Certainly, the cacti cannot survive indefinitely without water and nutrient supply from the substratum. In the strongly seasonal salinas, of course, they only need to overcome the dry season when salinity-stress is present, particularly if they are able to form functional roots again in the wet season. Indeed, cacti are known to be capable of rapid **adventitious root regeneration** (Fig. **6.14C**). *S. ottonis* also develops a large horizontal root system from which strong vertical tap roots protrude into the soil, and from which fine absorptive roots have a seasonal turnover related to substratum salinity (Fig. **6.16**).

To overcome an extended dry season, the insulated stems of the cacti use the possibility of nocturnal **recycling of respiratory CO₂** provided by the

Fig. 6.16A, B. Root system of *Subpilosocereus ottonis* as shown on a seedling (**A**) and a tall fallen cactus (**B**) with extended horizontal root system and vertical tap roots

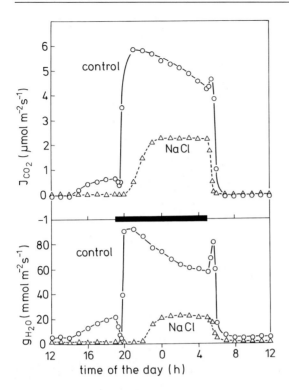

Fig. 6.17.
Effect of NaCl on photosynthetic gas exchange of small seedlings of *Cereus validus*. The control plants were irrigated with water; the NaCl-treated plants received NaCl solutions of daily increments of 50 mM until 400 mM NaCl was reached and were then kept for 16 days at this NaCl level. The *horizontal black bar* indicates the dark period. (Nobel et al. 1984)

CAM-mechanism (see Sect. **3.7.3** and box **3.11**). The experimental seedlings of *C. validus* under salt stress reduced photosynthetic gas exchange (Fig. **6.17**), and the contribution of nocturnal CO_2-recycling to total night-time malate accumulation increased from 20 % in the controls to 50 % in the NaCl-treated plants. In the extreme case of total insulation, stomata may close permanently during both day and night, reducing gas exchange to an absolute minimum. In this way the plants do not gain carbon, but they minimize loss. Carbon from respiratory CO_2 is recycled into malate during the night and, after decarboxylation of malate, in photosynthetic CO_2-fixation and carbohydrate synthesis through the Calvin cycle during the day. Thus, metabolism is maintained by respiratory and photosynthetic energy turnover, and the **only input** under these conditions is **light energy**, which keeps them alive.

Naturally, some loss of water vapour occurs via **cuticular transpiration** and leads to a gradual reduction of the water reserves in the **water-storage parenchyma** of the cactus stems (Fig. **6.18**) and a decline of vitality. Its has been shown that cacti survive when up to 54 % of tissue water content is lost, although any subsequent loss is lethal (Holthe and Szarek 1985). Thus, the chance of a small cactus to survive the salinity stress of the dry season is much smaller than that of a large cactus. A certain minimal biomass, with

Fig. 6.18.
Loss of fresh weight, i.e. loss of water, by small plants of
Subpilosocereus ottonis (about 0.3 m tall) derooted and
placed in full sun exposure in the dry season starting on
day 0. Errors are SD, n was 6 to 18. (Lüttge et al. 1989a)

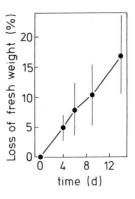

Fig. 6.19.
Profile of Cl⁻ levels in the soil of a large plant
of *Subpilosocereus ottonis* on a vegetation
island in the sand plain of Chichiriviche, Venezuela, during the dry season. (Medina et al.
1989)

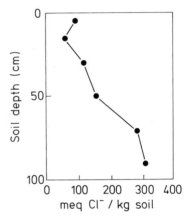

a sufficiently large water reserve in the water-storage parenchyma, is
required. Of course, survival also depends on the length of the dry and wet
season, respectively. If the wet season is longer and the dry season relatively
short, newly established seedlings have a better chance of survival. Thus, as
shown for cacti and agaves in the deserts of North America (Jordan and
Nobel 1979, 1982) seedlings do not survive every year, and one observes **age
classes** of larger plants, which indicates the wetter periods when seedlings
were able to become established.

For the larger cacti growing on the vegetation islands, salinity stress in
the top soil is reduced even in the dry season. The major tap roots of 6–7
m tall plants of *S. ottonis* are no longer than 50 cm, and even during the dry
season they do not extend below the point where salinity becomes 150
mequiv. Cl⁻ kg⁻¹ air-dried soil (Fig. **6.19**). In conclusion, the major problem
for the cacti in the salinas really is to survive the vulnerable seedling stage.

6.3.2.2
Tank-Forming Bromeliads

Some tank-forming terrestrial bromeliads occur on the salinas. At the northern coast of Venezuela by far the most frequent is *Bromelia humilis*, with the type II or **tank-root life form** (see Sect. **4.3.1** and Fig. **4.11B**). Since it does not necessarily need to form soil roots, the leaf rosettes may simply lie on the ground, and thus, *B. humilis* is effectively also a **salt excluder and stress avoider.**

Within tanks and through tank roots *B. humilis* can collect and utilize water. Therefore, in contrast to the columnar cacti (Sect. **6.3.2.1**), *B. humilis* can replenish water reserves even from very small spells of rain during the dry season. There is a marked peripheral water-storage parenchyma of thin-walled, non-green and highly vacuolated cells, with little cytoplasm at the adaxial surface of the leaves (see Fig. **4.18C**). The leaf tissue looses water during the dry season, and the leaves become less succulent and have increased dry weight : fresh weight ratios (Fig. **6.20**). Overall, the CAM plant *B. humilis* demonstrates **water storage at 3 different time scales:**

- short term storage based on the osmotic effects of nocturnal malate accumulation in the leaf cells (see Sect. **4.4.1**);
- medium term storage in the tanks;
- long term storage in the water parenchyma.

B. humilis occurs in different vegetation units on the sand plain and expresses three different **phenotypes** of growth form and pigmentation (see Sect. **3.6.2.2**), namely the dark green phenotye shaded under shrubs and trees of the deciduous forest, the yellow phenotype exposed on bare soil

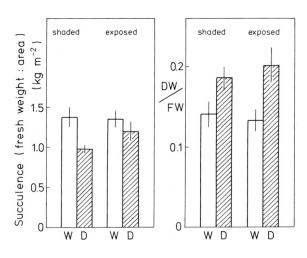

Fig. 6.20.
The degree of succulence (fresh weight : area) and dry weight : fresh weight (DW/FW) ratios in leaves of shaded and exposed plants of *Bromelia humilis* in the wet season (*W*) and in the dry season (*D*) in the sand plain of Chichiriviche, Venezuela. (Lee et al. 1989)

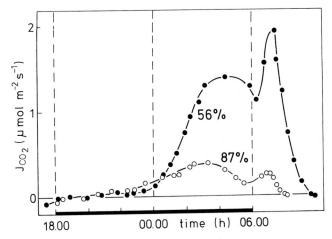

Fig. 6.21. Net CO_2 exchange (JCO_2) during the night and in the early morning of a green shaded (●) and a yellow exposed plant (■) of *Bromelia humilis* in the sand plain of Chichiriviche, Venezuela. The *horizontal black bar* indicates the night-period. (Lee et al. 1989)

islands or in the grassland of the sand plain, and a light-green intermediate phenotype, which also grows in relatively exposed conditions (Fig. **3.36**). The differential characteristics of shade and sun plants (Sect. **3.6.2.2**) are fully expressed in these phenotypes. In the dry season net CO_2 exchange was reduced with 7.2 mmol m^{-2} day^{-1} in the yellow and 22.2 mmol m^{-2} day^{-1} in the green and shaded plants (Fig. **6.21**), while an average rate obtained for all phenotypes in the rainy season was 33.0 mmol m^{-2} day^{-1}.

In the dry season the plants operated with increasing internal CO_2-**recycling** which corresponded to 56 % in the green and 87 % in the yellow phenotype (Fig. **6.21**), while recycling for all phenotypes in the wet season was only 21 %. Clearly, the yellow exposed plants are under the most severe stress of

- high irradiance,
- drought, and
- low nutrient supply.

The latter applies because the exposed plants also lack-supplies from decomposing litter falling between the plants and into the tanks of the shaded plants in the deciduous forest. This is reflected in their productivity. It is seen that losses and gains are more or less balanced and net productivity is close to zero in the yellow exposed plants, as compared to the green shaded plants which have a distinct primary net productivity (Table **6.1**).

Table 6.1. Data on productivity of exposed and shaded plants of *Bromelia humilis* in the alluvial plain of Chichiriviche, Venezuela. (Lee et al., 1989)

	Exposed	Shaded
Potential max. productivity (t DW ha^{-1} year^{-1})	6	6
Actual productivity (not including roots and ramets) (t DW ha^{-1} year^{-1})	0	3
FW/plant (kg)	0.18	0.54
Leaves/plant	42	70
New leaves/plant × year	36	48
Lost leaves/plant × year	36	23
New ramets/plant × year	0.24	0.67

Information on productivity was obtained from DW/FW determinations and repeated measurements and leaf counts of tagged plants over a period of 16 months from the end of one rainy season to the end of a dry season and again in the following year. DW = dry weight, FW = fresh weight.

6.3.3
Epiphytic CAM Plants: Avoidance of Salinity Stress

On the large cacti and the shrubs of the vegetation islands one frequently finds epiphytic CAM plants especially the bromeliad *Tillandsia flexuosa* and the orchid *Schomburgkia humboldtiana* (see Fig. **4.36**). Although there may be some salt spray driven inland by stronger winds, in their epiphytic habitat these plants largely avoid salinity stress. CAM serves adaptation to drought stress (Sect. **3.7.3**). Additionally both species are myrmecophilous (Sect. **4.4.2**). CO_2-acquisition in *S. humboldtiana* is greatly reduced in the dry season. In contrast, rates of CO_2-uptake are constantly low in *T. flexuosa* over both rainy and dry seasons (Fig. **6.22**). Internal CO_2-recycling is similar in both plants at 65–76 % and

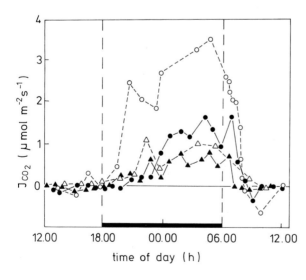

Fig. 6.22.
Net CO_2 exchange (J_{CO_2}) of *Schomburgkia humboldtiana* (●, o) and *Tillandsia flexuosa* (▲ Δ) in the wet season (o, Δ) and in the dry season (●, ▲). The *horizontal black bar* indicates the night period. (Griffiths et al. 1989)

independent of season. *T. flexuosa* is more frequent and so, despite lower potential maximum productivity, the physiological characteristics maintain carbon acquisition continuously over the seasons.

6.3.4
Mangroves and Associates

The shrubby vegetation of the islands is dominated by the mangrove associate *Conocarpus erectus*. The true mangrove *Avicennia germinans* also plays a limited role, particularly on vegetation islands closer to the lagoon with more permanent salt stress. *A. germinans* appears to be much more salt tolerant than *C. erectus*. The more detailed ecophysiological comportment of the two species is discussed in Sect. **5.4**.

6.4
Conclusions

As already discussed in Chap. **5**, salt stress has much in common with drought stress, in that water budget is the most important target of stress. Hence, it is no surprise that strategies recognised as an adaptation to limited water supply are also predominant among the plant species of the tropical salinas.

While the true halophytes are salt includers, and need compatible solutes in the cytoplasm when NaCl is sequestered in the vacuoles, other strategies involve leaf shedding and the use of the C_4- and CAM modes of photosynthesis. Among the plants of the tropical salinas CAM appears to be a typical strategy of salt excluders and salinity-stress avoiders. This may not be generalized, however, since CAM plays an important role in the non-tropical annual facultative halophyte *Mesembryanthemum crystallinum*, which is a strong salt includer when subject to salinity. However, again salinity and drought act in a similar way. Both serve to induce CAM in *M. crystallinum*, which also is a facultative CAM plant capable of switching from C_3-photosynthesis to CAM which extends its annual life cycle and allows to bring a large number of seeds to maturity (Lüttge 1993).

References

Griffiths H, Smith JAC, Lüttge U, Popp M, Cram WJ, Diaz M, Lee HSJ, Medina E, Schäfer C, Stimmel K-H (1989) Ecophysiology of xerophytic and halophytic vegetation of a coastal alluvial plain in northern Venezuela. IV. *Tillandsia flexuosa* Sw. and *Schomburgkia humboldtiana* Reichb., epiphytic CAM plants. New Phytol 111:273–282

Henriques RPB, Araújo DSD De, Hay JD (1986) Descrição e classificação dos tipos de vegetação da restinga de Carapebus, Rio de Janeiro. Rev Brasil Bot 9:173–189

Holthe PA, Szarek SR (1985) Physiological potential for survival of propagules of Crassulacean acid metabolism species. Plant Physiol 79:219–224

Jordan PW, Nobel PS (1979) Infrequent establishment of seedlings of *Agave deserti* (Agavaceae) in the northwestern Sonoran desert. Am J Bot 66:1079–1084

Jordan PW, Nobel PS (1982) Height distributions of two species of cacti in relation to rainfall, seedling establishment and growth. Bot Gaz 143:511–517

Lee HSJ, Lüttge U, Medina E, Smith JAC, Cram WJ, Diaz M, Griffiths H, Popp M, Schäfer C, Stimmel K-H, Thonke B (1989) Ecophysiology of xerophytic and halophytic vegetation of a coastal alluvial plain in northern Venezuela. III. *Bromelia humilis* Jacq., a terrestrial CAM bromeliad. New Phytol 111:253–271

Lüttge U (1993) The role of Crassulacean acid metabolism (CAM) in the adaptation of plants to salinity. New Phytol 125:59–71

Lüttge U, Medina E, Cram WJ, Lee HSJ, Popp M, Smith JAC (1989a) Ecophysiology of xerophytic and halophytic vegetation of a coastal alluvial plain in northern Venezuela. II. Cactaceae. New Phytol 111:245–251

Lüttge U, Popp M, Medina E, Cram WJ, Diaz M, Griffiths H, Lee HSJ, Schäfer C, Smith JAC, Stimmel K-H (1989b) Ecophysiology of xerophytic and halophytic vegetation of a coastal alluvial plain in northern Venezuela. V. The *Batis maritima – Sesuvium portulacastrum* vegetation unit. New Phytol 111:283–291

Medina E, Cram WJ, Lee HSJ, Lüttge U, Popp M, Smith JAC, Diaz M (1989) Ecophysiology of xerophytic and halophytic vegetation of a coastal alluvial plain in northern Venezuela. I. Site description and plant communities. New Phytol 111:233–243

Nobel PS, Lüttge U, Heuer S, Ball E (1984) Influence of applied NaCl on crassulacean acid metabolism and ionic levels in a cactus, *Cereus validus*. Plant Physiol 75:799–803

Smith JAC, Popp M, Lüttge U, Cram WJ, Diaz M, Griffiths H, Lee HSJ, Medina E, Schäfer C, Stimmel K-H, Thonke B (1989) Ecophysiology of xerophytic and halophytic vegetation of a coastal alluvial plain in northern Venezuela. VI. Water relations and gas exchange of mangroves. New Phytol 111:293–307

Walter H (1973) Die Vegetation der Erde in ökophysiologischer Betrachtung, vol 1. Die tropischen und subtropischen Zonen. G Fischer, Jena

Walter H, Breckle S-W (1984) Ökologie der Erde, vol. 2. Spezielle Ökologie der tropischen und subtropischen Zonen. G Fischer, Stuttgart

CHAPTER 7

Savannas

7.1
Physiognomy and Terminology

Savannas are open habitats typically dominated by grasses and often strongly affected by seasonal changes of rainfall. The term savanna, *sabana* in Spanish and *savana* (or campo) in Portuguese, is a West Indian expression of uncertain Caribbean origin (Huber 1987). A classical example of a savanna is the Llanos north of the Orinoco in Venezuela (Fig. 7.1). Alexander von Humboldt vividly described the seasonal contrasts in this large Venezuelan savanna-area:

"When under the vertical rays of the sun, never covered by clouds the combusted layer of grass has fallen into dust, the hardened soil cracks as if it were shaken by mighty earthquakes."[1]

"The uniform vision of these steppes has some greatness but also some tristesse and depression in it. It is as if the whole nature would be frozen; scarcely every now and then the shade of a small cloud, which hurries across the zenith and announces the near rainy season, falls over the savanna. One hardly can get used to the vision of the Llanos, which offer a picture like the surface of the sea..."[2]

"...as a saying goes here: 'The large ocean of greenery' ('los Llanos son como un mar de yerbas')"[2]

"When after a long drought the beneficial rainy season sets in, the scene in the steppe suddenly changes. When the surface of the earth is just wetted the fragrant steppe covers itself with *Kyllingias* with highly panicled *Paspalum* and with a diversity of grasses. Stimulated by the light, herbaceous mimosas open their folded dormant leaves and greet the rising sun like the early song of the birds."[3]

---

[1] "Wenn unter dem senkrechten Strahl der niebewölkten Sonne die verkohlte Grasdecke in Staub zerfallen ist, klafft der erhärtete Boden auf, als wäre er von mächtigen Erdstößen erschüttert..."

[2] "Der einförmige Anblick dieser Steppen hat etwas Großartiges, aber auch etwas Trauriges und Niederschlagendes. Es ist, als ob die ganze Natur erstarrt wäre, kaum daß hin und wieder der Schatten einer kleinen Wolke, die durch den Zenith eilend die nahe Regenzeit verkündet, auf die Savanne fällt. Nur schwer gewöhnt man sich an den Anblick der Llanos, die... ein Bild der Meeresflächen bieten."
"... wie man hier oft sagen hört, 'dem großen Meer von Grün'...'los llanos son como un mar de yerbas'..."
(Südamerikanische Reise, A. v. Humboldt 1808/1982)

[3] "Tritt endlich nach langer Dürre die wohltätige Regenzeit ein, so verändert sich plötzlich die Szene in der Steppe. Kaum ist die Oberfläche der Erde benetzt, so überzieht sich die duftende Steppe in Kyllingien, mit vielrispigem *Paspalum* und mit manigfaltigen Gräsern. Vom Lichte gereizt, entfalten krautige Mimosen ihre gesenkt schlummernden Blätter und begrüßen die aufgehende Sonne, wie der Frühgesang der Vögel."
(Ansichten der Natur, A. v. Humboldt 1849/1986)

Fig. 7.1A-C. The Llanos, Venezuela (**A**, **B**) and a savanna in Costa Rica, protected within the borders of the Santa Rosa Park (**C**)

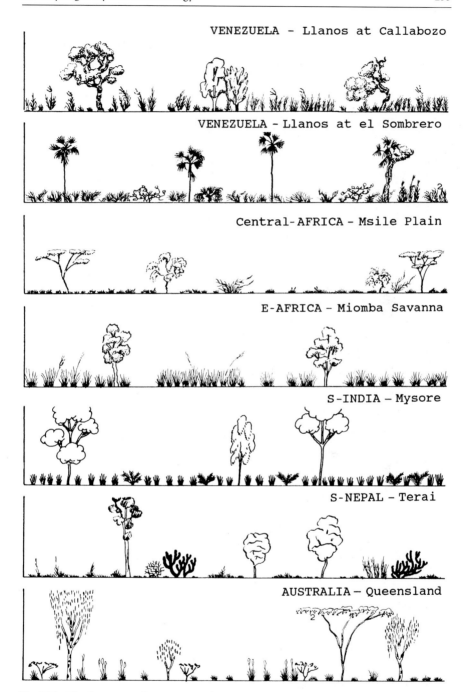

Fig. 7.2. Physiognomy of savannas with examples from all over the world. Transects of Vareschi (1980, with kind permission of R. Ulmer)

In order to give an impression of the vegetation Fig. **7.1** shows a picture of the Llanos in Venezuela and a savanna in Central America, and the drawings of transects in Fig. **7.2** present various types of savannas from all over the world. H. Walter (Walter and Breckle 1984) distinguished between savanna and park-land as follows:

- **Savanna:** homogeneous plant communities with scattered woody plants (trees, shrubs and bushes) in a grass layer closed in a greater or less degree over the soil and with a few herbs in between.
- **Park-land:** a mosaic of forest islands in an open grassland with few woody plants where the forest is associated with biotopes (e.g. river banks, valley bottoms, hills), which are ecologically different from the grassland.

In contrast, the geographer C. Troll (see Walter and Breckle 1984) considered various types of landscapes as savannas which constitute a macromosaic of grassland with different tree-formations (Fig. **7.3**). Some of them are illustrated in the accompanying photographs:

- the gallery forest along rivers (Fig. **7.4**),
- the forest of gullies,
- the termite savanna (Fig. **7.5**),
- the "morichales" (Fig. **7.6**).

This already illustrates the problems inherent in the use of the term savanna, which has been applied to a wide range of habitats.

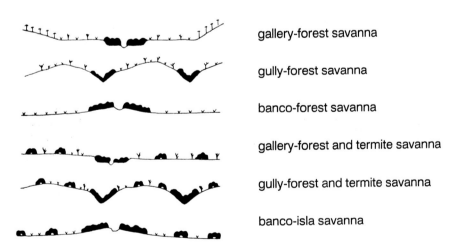

gallery-forest savanna

gully-forest savanna

banco-forest savanna

gallery-forest and termite savanna

gully-forest and termite savanna

banco-isla savanna

Fig. 7.3. Macromosaics of plant formations with grassland in landscapes called savannas by C. Troll. (Walter and Breckle 1984, with kind permission of S.-W. Breckle and G. Fischer-Verlag)

Fig. 7.4A, B. Gallery forests. **A** Rio Parupa, Gran Sabana, Venezuela. **B** In the cerrados near Brasília (Fazenda Agua Limpa), Brazil

Fig. 7.5A, B. Termite savannas. **A** After a fire, Queensland, Australia. The termite nests scattered over the field are well visible after the vegetation has been burnt. **B** *Acacia* wooded savanna, Great Rift Valley, Ethiopia

Fig. 7.6. Savanna with dense stands of the moriche palm, *Mauritia flexuosa*. Such morichales are restricted to wet marshy parts of savannas. (Gran Sabana, Venezuela.) Alexander von Humboldt has described the morichales as follows:
The palm *Mauritia flexuosa* "at moist places forms magnificent groups of fresh and shiny greenery, which recalls the green of our alder bushes. With their shade these trees maintain the moisture in the soil ..."[4]

Attempts to delineate the term more precisely have been made repeatedly (Huber 1982, 1987, 1990). Clearly, in the open habitats of the savannas the herbaceous ground-layer is the ecologically decisive stratum. There may be shrubs and trees, scattered or in small groups, but they never form a closed canopy. Thus, the major contribution to the **input of energy into the whole ecosystem** (by capture of solar radiation and primary biomass production) comes from the **herbaceaous ground-stratum**. The herbaceous layer itself may be dominated by grasses and sedges or by broad leaved herbs so that one can distinguish between

– **grass savannas** (Fig. 7.1), i.e. the savannas *sensu strictu*, and
– **herb savannas** (Fig. 7.7), which as "yerbazal" in Spanish, may not be considered as savannas in a strict sense but then would require use of a rather awkward term in English, e.g. "broad-leaved meadows".

---

[4] Sie bildet an feuchten Orten herrliche Gruppen von frischem glänzendem Grün, das an das Grün unserer Ellergebüsche erinnert. Durch ihren Schatten erhalten die Bäume die Nässe des Bodens ..."
(A. v. Humboldt, Ansichten der Natur 1849/1986.)

Fig. 7.7. *Yerbazal* (herb savanna), Sierrania Parú (04° 25'N, 65° 32'W, 1250 m a.s.l.), dominated by *Stegolepis hitchcockii* (broad flat leaves), *Brocchinia hechtioides* (slender tank-forming bromeliad) and *Bonnetia crassa* (small shrub)

Typical tropical grass savannas are restricted to low and medium elevations not exceeding 1000-1200 m, while herb savannas may occur at higher elevations. In either case it is essential that we are dealing with tropical or subtropical ecosystems (Fig. **1.3 A**), where in contrast to the tropical forests with a closed canopy of trees (Chap. **3**) **the ground stratum of grasses and/ or herbs dominates energy turnover** (Huber 1982, 1987, 1990). This may also occur in ecosystems outside the tropics, which then are distinguished as prairies, pampas or steppes. This strict distinction was not then familiar to Alexander von Humboldt who used "savanna" and "steppe" synonymously. On the other hand, the term savanna has also been used for physiognomic characterization of vegetation outside the tropics (Eiten 1986).

Even in the neighbouring countries of South America, Venezuela and Brazil, classification of savanna-like vegetation has led to different terminologies (Sarmiento 1984; Eiten 1972, respectively). For the vegetation in Brazil the **cerrado-concept** was developed, which is somewhat narrower than the

Table 7.1. Physiognomic description of various types of Venezuelan savannas and Brazilian cerrados. (After Sarmiento 1984; Eiten 1972)

	Terminology	
Description	Venezuela	Brazil
1. Savannas without woody species taller than the herbaceous stratum	Grass savanna	Campo limpo
2. Savannas with low woody species (< 8 m) forming a more or less open stratum		
a) Shrubs and trees isolated or in small groups, < 2 % of total surface	Tree and shrub savanna	Campo sujo
b) Shrubs and trees 2-15 % of total surface	Woodland or bush savanna	Campo cerrado
c) Trees > 15 % of total surface	Woodland	Cerrado *(sensu strictu)*
3. Savannas with tall trees (> 8 m)		
a) Isolated trees, < 2 % of total surface	Tall tree savanna	
b) Trees 2-15 % of total surface	Tall tree savanna woodland	
c) Trees 15-30 % of total surface	Tall tree wooded grassland	
d) Trees > 30 % of total surface	Tall woodland	Cerradão
4. Savannas with large trees in small groups	Park savanna	Campo coperto
5. Mosaic of units of savannas and forests	Park	

Fig. 7.8A-D. Aspects of the cerrado (**A**, **B**) and the cerradão (**C**, **D**) near Brasília (Fazenda Agua Limpa), Brazil

more general **savanna-concept.** In Brazil, 20 % of the area of the whole country and 40 % of the non-Amazonian part are covered by cerrados. They are geographically and ecologically intermediate between tropical rainforest and tropical/subtropical desert. The annual precipitation averages between 1200 and 1600 mm ranging from 800 mm to 2000 mm in the driest and wettest parts, respectively. There is strong seasonality with more than 90 % of the rain falling in 7 months (October-April). The cerrado soil is very infertile and can vary from less than 5 % to over 95 % sand, the rest is clay and a little silt (Eiten 1972, 1986).

Table **7.1** attempts a systematic comparison of the terms used to describe Venezuelan and Brasilan savannas and cerrados (see also Figs. **7.1** and **7.8**). The density and size of woody plants, i.e. shrubs and trees is an important feature of this system. In Africa one also encounters the distinction between "wooded savanna", where the trees stand more or less isolated, and "woodland savanna," where the canopies of individual trees touch each other (Fig. **7.9**).

A more fundamental problem is why there are savannas at all. Why is closed forest not growing all over these sites in the tropics? Are savannas natural plant communities or only products of human activities? There is no generally accepted hypothesis, and a number of possibilities are listed by Huber (1982) as follows:

- **The climatic hypothesis.** This must be rejected for several reasons, but most simply because of the co-occurrence of forest with closed canopies of trees and open savanna under the same climatic conditions.

- **The edaphic hypothesis**, especially including the importance of the nutrient limitation.

- **The fluvial hypothesis**, i.e. colonization of ancient riverbeds by savanna.

- **The hydrological hypothesis**, i.e. the important influence of the water regime including limitations due to insufficient or excessive drainage.

- **The relict or refuge hypothesis**, where savannas are considered as relicts of a formerly more widespread dry vegetation type.

- **The anthropogenic hypothesis**, implying the role of man in establishing, maintaining and extending savannas especially by forest clearing and burning.

Clearly, savannas are not only man made. The cerrado of central Brazil also is a natural, original vegetation and not derived from tropical mesophytic forest by man's destruction. Cerrados and natural savannas in the tropics are highly valuable biotopes both floristically and ecologically. It is mostly

Fig. 7.9. A Wooded savanna, **B** woodland savanna, Great Rift Valley, Ethiopia

overlooked, that they are just as much threatened by the current destruction as tropical forests. There may even arise some kind of unsavoury contest in that destruction will increasingly turn towards savannas as forests are protected.

The major factors which determine savannas are:

- water (see above: hydrological hypothesis),
- mineral nutrients (see above: edaphic hypothesis),
- herbivory (cattle, see above: anthropogenic hypothesis) and
- fire

(Högberg 1986a), some of which are discussed in the subsequent sections.

7.2
The Water Factor

According to water status and seasonality in his **climatic-hydrological classification** Sarmiento (1984) distinguishes mainly four types of savannas based on water status and seasonality:

– a **semi-seasonal savanna** with a long rainy period but without excess of water (i.e. flooding) and a short period with a water deficit;

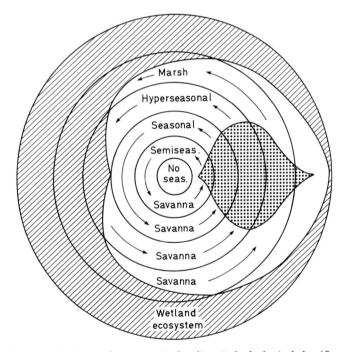

Fig. 7.10. Scheme of the water budgets of savannas in the climatic-hydrological classification of Sarmiento (1984). Following the annual cycle described by the circumference of each circle shows the extensions of annual cycles dominated by water excess (*hatched area*), normal water availability (*white area*) and water deficit (*dotted area*) (Reprinted by permission of Harvard University Press)

Box 7.1

Model of the water budget of savannas with a continuous vegetation-soil compartment separated from the atmosphere. Rom. = losses, ital. = input. (Sarmiento 1984; reprinted by permission of Harvard University Press).

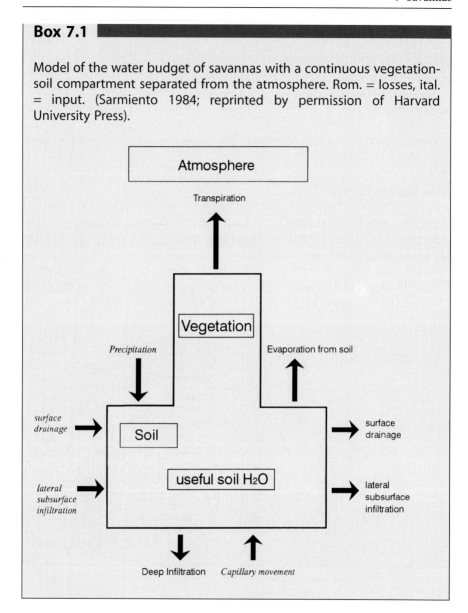

- a **seasonal savanna** with changes between periods with sufficient water and periods of drought;
- a **hyperseasonal savanna**, where periods of excess of water and of drought provide strong seasonal contrast (see quotations of A. von Humboldt in Sect. **7.1**);
- a **marsh savanna**, where long periods of water excess are interrupted by drier periods, when water, however, still is in sufficient supply.

Table 7.2. Water relations of two types of savannas compared to a tropical rainforest. (Sarmiento 1984)

	Precipitation (mm p.a.)	Evapotranspiration (mm p.a.)	Free evaporation (mm p.a.)
Trachypogon savanna (Calabozo, Venezuela)	1839	1440	2406
Leptocoryphium savanna (Barinas, Venezuela)	1170	1115	2156
Rainforest (Ivory Coast)	1800 – 1950	965 – 1000	[a]

[a] Infiltration deeper than 2.3 m: > 600 mm p.a. (in savannas = 0).

These relationships are shown schematically in Fig. **7.10**. A model of the water budget of savannas explaining the various inputs and losses is shown in Box **7.1**. The key elements, of course, are precipitation for the input and evapotranspiration by the vegetation plus free evaporation for the losses. The comparison between two types of savannas and a rainforest in Table **7.2** then suggests that the main differences between the savannas and the forest are the very high free evaporation and the deep infiltration, respectively.

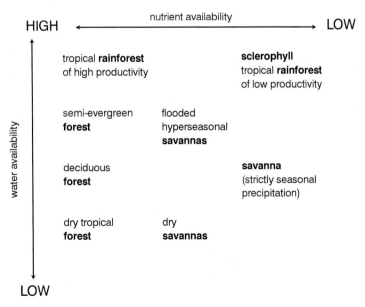

Fig. 7.11. Separation of various types of savannas and tropical forests based on nutrient and water availability. (Medina 1987).

Table 7.3. Different requirements of grasses and woody plants in savannas with respect to the water factor. (After Walter and Breckle 1984)

	Grasses	Woody plants
(i)	*Amount of precipitation*	
	Lower amounts of **annual precipitation**	**Larger** amounts of **annual precipiation**
(ii)	*Annual distribution of precipitation*	
	Precipitation must occur during the **growth period**	**Precipitation** may occur during the **rest period**
(iii)	*Annual distribution of water uptake*	
	During the **rest period no water is taken up** from the soil	The soil must provide enough water for a minimal **water-uptake also in the dry season**
(iv)	*Soil-water capacity*	
	The **water capacity of the soil** must be **high:** the plants do not limit their transpiration as long as the soil provides enough water, and then the leaves dry rapidly	The **water capacity of the soil** does **not** need to be **high:** the crumb size of the soil may be large, the soil may be stony and rocky; the root system develops far in both horizontal and vertical direction

If nutrient supply is taken as a second dimension in addition to water supply, the two-dimensional separation of various types of tropical forests and savannas emerges as shown in Fig. **7.11**. The forests require high nutrient supply, or at least high availability of water when nutrient supply is small, as in sclerophyll forest and in low-productivity rainforest. Conversely, savannas occupy areas with medium to low supply of the two resources.

Woody plants and grasses in savannas have different requirements from the annual precipitation, dependent on the distribution of rainfall, soil availability of water over the year and the water-capacity of the soil (Table 7.3). It is therefore necessary to consider grasses and trees separately.

7.2.1
Grasses

The following **phenological groups** are observed among savanna grasses:

- perennial with a seasonal semi-dormant-period;
- annual, ephemeral with a short cycle;
- annual with a long cycle;
- perennial with a seasonal dormant period;
- continuous growth and flowering.

Fig. 7.12 A-C.
Semiquantitative pheno-
grams (relative units on
the ordinates) of two tropical grasses with C_4 photosynthesis (**A**, **B**) and a
tropical savanna tree (**C**)
(Sarmiento 1984; reprinted
by permission of Harvard
University Press)

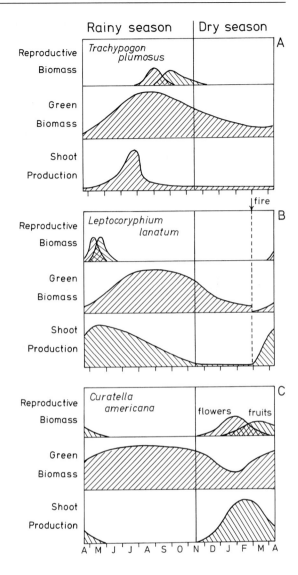

The first group is most frequent. It is represented, for example, by *Trachypo-gon plumosus* and *Leptocoryphium lanatum* (Fig. **7.12**). The phenological diagrams for the two species, in contrast to the tree *Curatella americana*, Dilleniaceae (see below Sect. **7.2.2**), show flowering and new shoot produc-tion during the rainy season, but closer to the end of the rainy season in *T. plumosus* than at the beginning as for *L. lanatum* (Fig. **7.12**). The phenolog-ical diagram of *L. lanatum* also shows the stimulation by fire of shoot pro-duction at the end of the dry season (see below Sect. **7.4** for more details). Such phenological diagrams, as discussed by Solbrig (1993) allow the sepa-

Box 7.2

Biochemical pathways of C₄ plants

Primary CO_2 fixation via phosphoenolpyruvate carboxylase (**PEPC**) and refixation of CO_2 via ribulose-bis-phosphate carboxylase (**RUBISCO**) occur **simultaneously in time** and are **separated in space**.

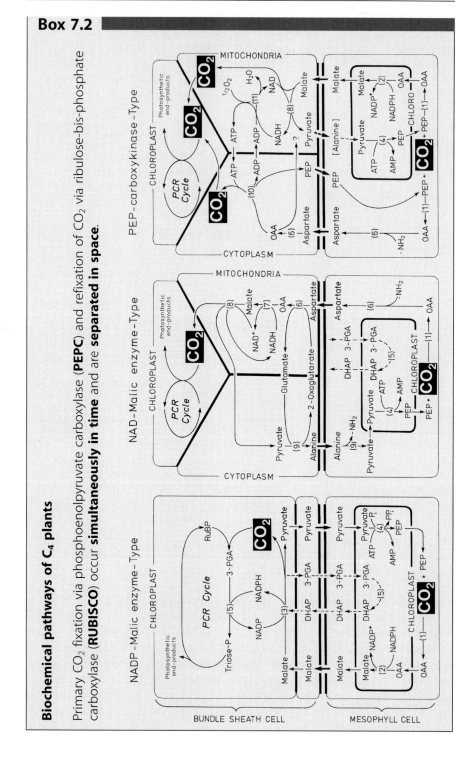

Box 7.2 (Continued)

The biochemical reactions in the mesophyll are basically similar in all types of C_4 plants. The first CO_2 fixation product is the C_4 acid anion oxaloacetate, which is subsequently transformed to malate and/or aspartate. Three types of C_4 plants are distinguished by the mode of decarboxylation of these C_4 acids after their transport to the bundle-sheath cells:

- the NADP-malic enzyme type [reaction (3)],
- the NAD-malic enzyme type [reaction (8)],
- the PEP-carboxykinase type [reaction (10)].

Enzymatic reactions
(1) PEP-carboxylase (PEPC),
(2) NADP-dependent malate dehydrogenase,
(3) NADP-dependent malic enzyme,
(4) Pyruvate, P_i dikinase,
(5) 3-PGA kinase, NADP-dependent glyceraldehyde-3-P dehydrogenase and triose-P isomerase,
(6) Aspartate aminotransferase,
(7) NAD-dependent malate dehydrogenase,
(8) NAD-dependent malic enzyme,
(9) Alanine aminotransferase,
(10) PEP-carboxykinase,
(11) Mitochondrial NADH oxidation systems.

Metabolites and cofactors
AMP, ADP, ATP: adenosine mono-, di- and tri-phosphate;
DHAP: dihydroxyacetone phosphate;
NAD: nicotine-adenine-dinucleotide;
NADP: nicotine-adenine-dinucleotide phosphate;
OAA: oxaloacetic acid;
P: phosphate;
PCR: photosynthetic carbon reduction;
P_i: inorganic phosphate;
PP_i: inorganic pyrophosphate;
PEP: phosphoenolpyruvate;
PGA: phosphoglyceric acid;
RubP: ribulose-bis-phosphate.

(Hatch and Osmond 1976; Hatch 1987)

ration of functional groups of savanna grasses. Thus, there are grasses that grow early and reproduce quickly in the rainy season (more like *L. lanatum* in Fig. **7.12**) as opposed to grasses that grow gradually and develop shoots slowly and reproduce in the middle or towards the end of the rainy season (more like *T. plumosus* in Fig. **7.12**). In general terms, the former (i.e. early growers and reproducers)

- are more drought resistant,
- have higher turgor pressures during the dry season,
- have higher water use efficiencies (WUE),
- partition more of their photosynthesic products to roots and below-ground organs,
- are more competitive under dry conditions and have increased importance along wet to dry gradients.

Grasses need less water than savanna trees, but the water must be available during the growth period.

A major ecophysiological aspect of savanna grasses is the dominance of **C$_4$-photosynthesis** (Box **7.2**). Somewhat similar to CAM (see Sect. **3.7.3** and Box **3.11**), during C$_4$-photosynthesis CO$_2$ is first cycled through malate, before it is assimilated in the Calvin cycle. However, during C$_4$ photosynthesis there is simultaneously CO$_2$-fixation via PEP-carboxylase (PEPC), together with CO$_2$-remobilisation, refixation via RUBISCO and reduction via the Calvin cycle. The PEPC- and RUBISCO-functions are localized in different cell types and hence spearated spatially, in contrast to CAM, where they occur in the same cells but are separated in time. Leaves of C$_4$-plants have two distinct photosynthetic tissues, first the outer spongy mesophyll where CO$_2$ is fixed to produce malate and in many cases also aspartate via oxaloacetate, and second, the inner bundle sheath where malate or aspartate are decarboxylated and the CO$_2$ is refixed. Depending on the enzymic mechanism of decarboxylation, 3 different types of C$_4$-photosynthesis can be distinguished (Box. **7.2**). Because the affinity of PEPC for CO$_2$ is about 60times higher than that of RUBISCO, fixation of atmospheric CO$_2$ in the mesophyll, which tightly surrounds the bundle sheath, is highly effective. The malate and aspartate so produced are transported symplastically to the bundle sheath, via plasmodesmata connecting the two tissues. Frequently, there is suberization of the cell walls between the two tissues to prevent leakage to the apoplast, and decarboxylation in the bundle sheath leads to a 6-10 fold increase of CO$_2$-concentration as compared to atmospheric CO$_2$. This has several ecophysiological advantages which are important for savanna grasses in dry open habitats with high irradiation:

• Under water stress the high CO$_2$-affinity of the first step of CO$_2$-fixation (PEPC) draws down CO$_2$-concentration inside the leaf, providing a steeper gradient for inward diffusion of CO$_2$, and allows **operation of photo-**

Box 7.3

Some major ecophysiological characteristics of the three major modes of photosynthesis in terrestrial plants, viz. CAM (*D* dark; *L* light period), C_4 and C_3 photosynthesis. (Black 1973)

	CAM	C_4	C_3
Water-use efficiency (WUE) (mol [CH_2O]: mol H_2O)	$(6-30) \times 10^{-3}$(D) $(1-4) \times 10^{-3}$(L)	$(1.7-2.4) \times 10^{-3}$	$(0.6-1.3) \times 10^{-3}$
Maximum net CO_2 uptake ($\mu mol\ m^{-2}\ s^{-1}$)	0.5–2.5 (D) 7–8 (L)	25–50	10–25
Maximum productivity ($g\ DW\ m^{-2}\ d^{-1}$)	1.5–1.8	400–500	50–200

synthesis with partially closed stomata, which reduces transpiratory loss of H_2O. Hence, the **water use efficiency** of C_4-plants is much higher than that of C_3-plants, although still lower than that of nocturnal CO_2-fixation in CAM plants (Box **7.3**).

- The high CO_2-affinity of PEPC, together with the simultaneous use of light in refixation of CO_2 via RUBISCO, also allows **high maximum rates of photosynthesis** and high productivity, which in C_4-plants are the highest of the three modes of photosynthesis (Box **7.3**).
- The **high CO_2-concentration in the bundle sheath cells** reduces photorespiration and the plants are less susceptible to the danger of photoinhibition and photodamage.

Therefore, it is not surprising that C4-grasses dominate in tropical savannas and C_3-grasses are more scarce. The most frequent tribes and genera are listed in Table **7.4**, and Table **7.5** gives representative rates of photosynthesis with the highest maximum rates obtained in the C_4-grasses. Some tropical C_4-grasses have been developed into major agricultural crops of mankind, e.g. sorghum, millet, maize and sugar cane; the Ethiopian tef is also a C_4-grass, *Agrostis tef*.

African C_4-grasses have been introduced and are invading savannas in the neotropics and displacing the native herbaceous vegetation. In relation to the advantages of C_4-photosynthesis listed above the competitive superiority of the African grasses is due to

- higher net-photosynthesis rates,
- more efficient use of soil nutrients,
- higher proportion of assimilates allocated to new leaves,
- higher tolerance to defoliation.

Table 7.4. Genera of neotropical savanna grasses according to their photosynthetic pathway. (Reference to Venezuelan tribes, Montaldo 1977; Medina 1982)

Tribe	Genera
	C_4 grasses
Eragrosteae	Eragrostis, Leptochloa
Chlorideae	Microchloa, Bouteloua, Chloris, Gymnopogon
Sporoboleae	Sporobolus
Paniceae	Digitaria, Eriochloa, Paspalum, Echinochloa, Axonopus
Andropogoneae	Imperata, Andropogon, Trachypogon, Diectomis
Aristideae	Aristida
Arundinelleae	Tristachya
	C_3 grasses
Paniceae	Lasiacis, Oplismenus

Table 7.5. Representative rates of photosynthesis (μmol CO_2 m^{-2}s^{-1}) of savanna grasses with C_3- and C_4 photosynthesis. (Medina 1986)

	South American grasses	African grasses
C_3 grasses:	11.1 – 13.2	–
C_4 grasses:		
Field data	3.2 – 16.4	2.2 – 7.9
Laboratory data	28.1	14.7 – 43.9

Many grasses in the Llanos of Venezuela are also C_4-grasses. In the Llanos, the native grass *Trachypogon plumosus* and the successful invader *Hyparrhenia rufa*, both C_4-plants, are distinguished as follows:

- *H. rufa*, has higher transpiration and stomatal conductance for water vapour, using water opportunistically when available; it shows earlier leaf-senescence during the dry season, i.e. drought avoidance; however, it needs relatively deep soils;

- *T. plumosus*, uses water conservatively; it is more drought tolerant and can withstand poorer nutrient and water status on shallower soils (Baruch and Fernandez 1993).

Some work has also been devoted to the **altitudinal distribution of C_4 and C_3 plants in the tropics.** The determining factors are

- temperature,
- water availability,
- irradiance, and
- to a more limited extent nutrient availability.

Table 7.6. Altitudinal distribution of C_4- and C_3 plants in the tropics. The soil moisture index for Kenya is in arbitrary units increasing approximately linearly with altitude (10 at 600 m; 100 at > 3500 m)

Altitude (m)	Soil moisture index	C_4 vs. C_3 photosynthesis
Grasses in Kenya[a]:		
< 2000	50	C_4 grasses dominate
		C_3 grasses only in the shade
2000–3000	60–70	Transition zone
> 3000	80	C_3 grasses dominate
107 species from 11 families in NW central America [b]:		
0		40 % C_4 species
2600		< 6 % C_4 species

[a] Tieszen et al. 1979. [b] Meinzer 1978.

Although irradiance increases with altitude, temperatures are lower and water availability is larger. Therefore, C_4-grasses only dominate at lower altitudes and are replaced by C_3-grasses at higher altitudes with increasing soil-moisture index (Table **7.6**). CAM plants are rare in typical savannas. Stem succulent CAM species, i.e. Cactaceae in America and Euphorbiaceae in Africa, are important in dry thornbush-forests of tropical lowlands (Sect. **3.7.1**); CAM-epiphytes are abundant in wet forests of medium altitudes (Sects. **4.4.1** and **4.4.3**); and at higher altitudes, terrestrial CAM species may also be frequent in dry sites of open habitats (Sect. **9.3.3**).

Another site where C_4-photosynthesis has proved to be highly successful, is the fertile **flood plains** of nutrient rich rivers and lakes (white waters) of the **Amazon region** in South America. Here the perennial C_4-grass *Echinochloa polystachya* may form monotypic stands and displays extraordinarily high rates of net-CO_2 uptake in photosynthesis of 30–40 μmol m^{-2} s^{-1} (compare Table **7.5**) with fast growth and high productivity during the wet season when the flood plains are submerged. During the shorter dry period CO_2-uptake rates are 17 μmol m^{-2} s^{-1} and photosynthesis shows a midday depression (Piedale et al. 1994).

Based on the different carbon isotope ratios ($\delta^{13}C$) in C_3- and C_4-plants (see Sect. **2.5**) horizontal and vertical $\delta^{13}C$-analyses in soils have allowed historical studies of the dynamics of savanna-forest interfaces to be performed (Mariotti and Peterschmitt 1994). Where C_4-grasses dominate in savannas, the soil-organic matter beneath, fed by the decomposing litter of savanna plants, should be much less negative than the soil under forest, where C_3-trees dominate. Fig. **7.13** shows that at a study site in the Western Ghâts, India, deeper soil layers at 80 cm and below had more negative $\delta^{13}C$ values indicating that the whole area once was dominated by forest. The soil directly under the present savanna shows the less negative $\delta^{13}C$ values expected from the predominant C_4-photosynthesis by the vegetation and that under the present forest corroborates prevailing C_3-photosynthesis by

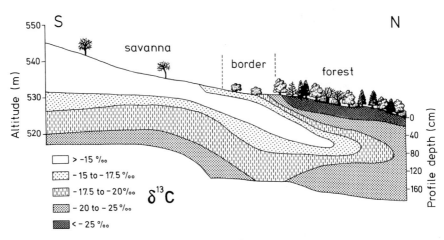

Fig. 7.13. Vertical and horizontal profiles of $\delta^{13}C$ values in the soil organic matter along a savanna-forest topological gradient in Kattinkar, Western Ghâts (13°57'N, 77°44'E), India. (After Mariotti and Peterschmitt 1994)

the forest trees. Interestingly, in the zone between forest and savanna, less negative $\delta^{13}C$ values extend deep under the present forest, indicating that the savanna must have had a larger extension in the past, and that the forest must be currently expanding.

7.2.2
Trees

Trees need larger amounts of precipitation than grasses. The requisite rain can fall also during dormant periods, and so a small amount of water uptake into the trunk and branches must be possible even during the drought periods. However, the water capacity of the soil does not need to be high. The trees can develop large root systems reaching far into the soil in both vertical and horizontal direction (Table 7.3).

In the Brazilian cerrados the soil is always deep and well drained. The ground-water table therefore is low, i.e. from 3-6 m down to 30-50 m. Hence, the trees develop **deep roots**, which reach water even during the dry season, as shown by high transpiration rates. In the Llanos in central Venezuela there is frequently a hard pan – "arecife" – of lateritic iron-oxide (see below Sect. 7.3.4) above the ground water-table (Fig. 7.14). The roots of savanna trees must penetrate this layer to reach the ground water, which also varies on a seasonal basis.

In addition trees develop xeromorphic structures which confer drought resistance. Some trees particularly have thickened **stems**, looking very **"succulent"** (Fig. 7.15) such as the Bombacaceae *Pseudobombax* in South America and *Adansonia*, the baobab. The latter is a most spectacular plant. There

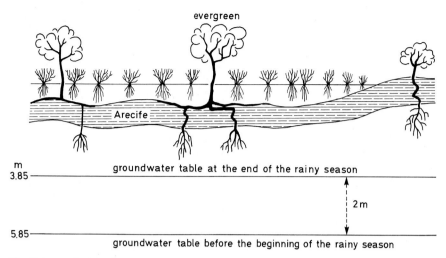

Fig. 7.14. Relations between the vegetation, the hard lateric ferrous-oxide layer ("arecife") and the seasonally shifted groundwater table in the Llanos of central Venezuela. (Walter and Breckle 1984, with kind permission of S.-W. Breckle and G. Fischer-Verlag)

is only one species on the African continent, *A. digitata*, which may be up to 9 m in diameter (Fig. **7.15D, E**) and has a geographical distribution clearly correlated with the occurrence of savanna (Fig. **7.16**). There are seven species of *Adansonia* in Madagascar and two in Australia. These "suc-

Fig. 7.15A (Continued on page 276)

Fig. 7.15A-E. "Stem-succulent" trees. **A** Caatinga in Brazil (Martius' Flora Brasiliensis, 1840-1906). **B** Savanna in Queensland Australia. **C** *Pseudobombax pilosus*, Paraguana Peninsula, Venezuela. **D, E** *Adansonia digitata* (baobab), Okawango Delta, March 1982, Botswana, Africa. (**D, E** courtesy Helga and Bodo Lüttge, Munich)

culent" stems are frequently referred to as water-stores. However, little work seems to be available on the putative ecophysiological role and function of water storage in these remarkable tree stems. Stable isotope ratios of *A. gregorii* in Australia were found to be $\delta^{13}C$ = -29.06‰ (indicating C_3-photosynthesis) and δD = -90.10‰ (H. Ziegler, unpubl.).

Fig. 7.16.
Geographical distribution
of the baobab in Africa.
(Walter and Breckle 1984,
with kind permission of
S.-W. Breckle and
G. Fischer-Verlag)

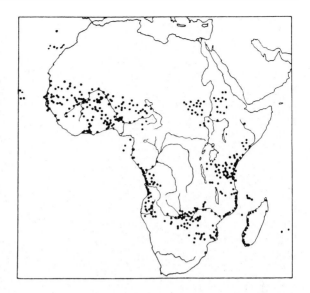

More generally, leaf xeromorphy is also observed among savanna trees. It is very important in Australia and South America, but less so in Africa (Medina 1993). It has already been noted in Sect. **3.5** that the formation of small and longlived leathery leaves may be considered as a strategy which gives the best return for investment of resources when nutrient supply is low. In addition such leaves also offer ways to economise on water use by some of the following traits:

- **dense venation;**
- **water storage tissues** (see also Fig. **4.12**);
- **thick and water tight cuticle**, which reduces water loss via cuticular transpiration;
- **dead hairs** on the surface;
- **sunken stomata;**
- prevailingly **hypostomatic** distribution of stomata , i.e. stomata only on the lower surface;

(the latter three properties are generally assumed to reduce evapotranspiration by supporting the built up of unstirred layers although in detail leaf boundary-layer relations are very complex, Schuepp 1993);

- **thick cell walls;**
- **lignification** of cell walls;
- formation of idioblasts and **sclereids**;

(these three properties help to stiffen the leaves, so that leaf-shape is maintained even when turgor pressure is low);

Fig. 7.17A-C

Fig. 7.17A-F.
Flowering savanna trees of the Llanos in Venezuela: *Tabebuia chrysantha* (**A**), *Tabebuia orinocensis* (**B**), *Yacaranda filicifolia* (**C**), *Pseudobombax* sp. (**D**), *Palicourea rigida* (**E**), *Byrsonima crassifolia* (**F**)

• production of **etheric oils**, which due to their hydrophobic nature may also assist in preventing water loss into the gas phase around the leaves.

Due to the strong seasonality of water supply in most savannas, appropriate **phenological behaviour** is also important. One of the most striking aspects

Fig. 7.18A, B.
Byrsonima crassifolia (**A**) and *Curatella americana* (**B**); Llanos, Venezuela, March 1991 (**A**), January 1989 (**B**)

is the attractive flowering of their trees (Fig. **7.17**). In Africa and Australia a few trees are evergreen and most are drought-deciduous, whereas in South America evergreen trees prevail (Medina 1993). *Curatella americana,* a dominant tree of the Venezuelan Llanos, is evergreen with seasonal growth. Production of flowers and fruits, new leaves and shoots begins in the middle to the last third of the dry season, so that the plants are "ready" when the rainy season comes, as shown in the phenological diagram of Fig. **7.12C**.

The **photosynthetic rates** of the mature leaves are extraordinarily high. For the two dominant trees of the Llanos (Fig. **7.18**) Medina (1982) records the following maximum rates of CO_2-uptake (μmol CO_2 m^{-2} s^{-1}):

Curatella americana	(Dilleniaceae)
in the laboratory	39
in the field	37
Byrsonima crassifolia	(Malpighiaceae)
in the laboratory	43

These compare with the average rates of 10-25 μmol m^{-2} s^{-1} for C_3-photosynthesis (see Box **7.3**). A **midday depression** (see Sect. **3.7.2**) may assist in regulating the water economy on hot days both during the rainy and the dry season (Fig. **7.19**).

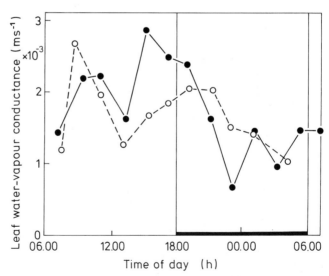

Fig. 7.19. Day-night cycle of leaf water-vapour conductance (g_{H_2O}) of the cerrado tree *Terminalia argentea* in Brazil showing a pronounced midday depression of the degree of stomatal opening both in the rainy season (o) and in the dry season (●). Note that for this experiment g_{H_2O} is given in the older units of m s^{-1} in contrast to the usage elsewhere in this volume. (Medina 1982)

Among the deciduous trees the length of the leafless periods may be different:

- trees with a short leafless period afford low (highly negative) water potentials and high respiration rates, e.g. *Bursera simaruba* (Burseraceae), *Spondias lutea* (Anacardiaceae), *Pereskia guamacho* (Cactaceae);

- trees with a long leafless period have high (less negative) water potentials and low respiration rates, e.g. *Tabebuia chrysantha* (Bignoniaceae).

The precise phenological behaviour of plants in the tropical savannas (see also grasses, Sect. **7.2.1**) is a reliable indicator of season, and raises the question of how this is sensed. Using up the last water reserves for formation of reproductive and vegetative biomass in the last third of the dry season would be dangerous if the rainy season were not close. In fact it has been observed that several phenological phenomena, including water budget, leaf abscission and flowering are related to phytochrome equilibria (see Box **3.2**) (Reich and Borchert 1984). Several suggestions have been made to explain

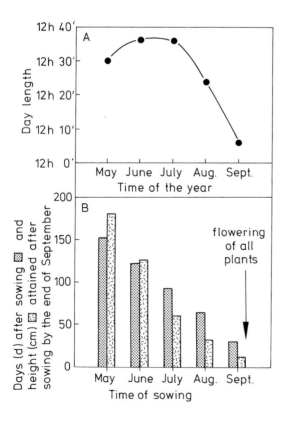

Fig. 7.20 A, B.
Experiments with the annual tropical short-day plant *Hyptis suaveolens* showing that plants can sense very small differences of photoperiod (daylength). Note that, by definition, short-day plants require daylengths below a certain species-specific threshold for induction of flowering. The daylengths from May to September at the site of the experiment are given in **A**. The columns in **B** give the number of days passed after sowing and the height of the plants attained after sowing by the end of September. Independent of both time passed after sowing and height attained, all plants flower at the end of September, i.e. the photoperiod of 12 h 24 min in August was still too long (above the threshold) but the only 18 min shorter photoperiod of 12 h 06 min in September was short enough to induce flowering in *H. suaveolens* at the tropical site. (After Medina 1982)

phenological timing on a hormonal basis. Water stress itself could be involved. Endogenous annual rhythms may play a role (Wright 1991). The proposal that the decisive external signal is **photoperiod** has been generally rejected, because near the equator the differences between the longest and the shortest days are rather small (Reich and Borchert 1984). However, the forest tree *Hildegardia barteri* in Nigeria occurs at 7 °N where photoperiod changes by 53 min, and a similar decrease of photoperiod inhibits seedling leaf production (Njoku 1963; Wright 1996). An experiment with *Hyptis suaveolens* (Lamiaceae), although an annual species and not a tree, really puts the debate in a different perspective (Medina 1982). It showed, that in principle plants can sense differences in photoperiod around 15 minutes (Fig. **7.20**). Seeds were germinated at the beginning of each month at a location north of the equator starting in May and ending in September. At the end of September, all plants were flowering irrespective of the age and biomass they had attained during growth, such that the 180 cm tall, ~150 day old plants, germinated in May flowered simultaneously with the 12 cm high, ~30 day old plants, only germinated in early September. Thus, flowering was not related to age or biomass. The photoperiod, which did not lead to flowering (plants germinated in August with no flowering in August photoperiod 12 h 24 min), and that which elicited flowering in September (photoperiod $\sim$ 12 h 06 min) differed by only $\sim$ 18 min. Thus, it appears, that many of the phenological phenomena observed in savannas in fact might well be regulated by photoperiod.

7.3
The Nutrient Factor

Mineral-nutrient relations of savannas are much determined by the **physicochemical properties of the upper soil layers,** such as

– texture,
– pH,
– cation exchange capacity,
– extractable bases (K^+ + Ca^{2+} + Mg^{2+} + Na^+, see also Sect. **7.5**)
– content of potentially mineralizeable N,
– availability of P

(Medina 1993). An important response of plants is the development of extended root systems and a symbiosis with fungi, namely **mycorrhiza.** Savanna soils are mostly very poor and infertile. The height and density of the woody layer in the cerrados (see Table **7.1**) depends on the fertility, depth and drainage of the soil and not on rainfall (Eiten 1972, 1986). The major nutrients limiting productivity of savannas are nitrogen and phosphorus, and an important stress factor is aluminium toxicity.

Box 7.4

Nitrogen cycles in ecosystems

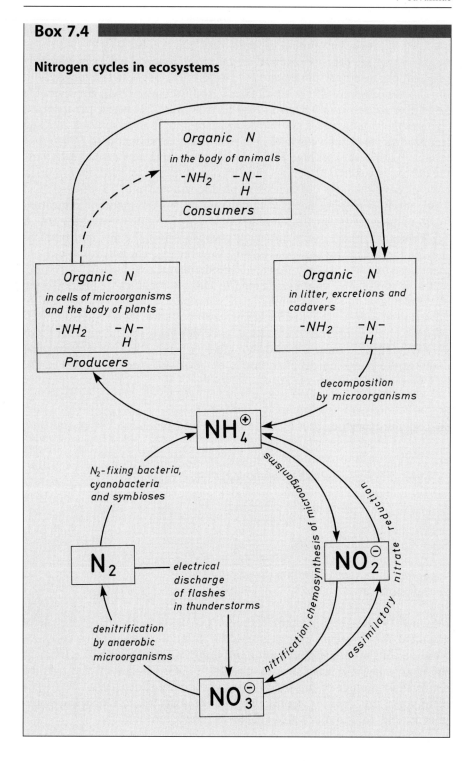

Box 7.4 (Continued)

Pools of N in various groups of organisms of the ecosystem and in various N-compounds (*boxes*) and transfer between the pools (*arrows*).

-NH_2, -N-, organic N;
NH_4^+, ammonia;
NO_2^-, nitrite;
NO_3^-, nitrate;
N_2, atmospheric dinitrogen gas.

(After Lüttge et al. 1994)

7.3.1
Nitrogen

7.3.1.1
Nitrogen Cycles

Nitrogen is one of the most critical elements for plant growth in savannas. The nitrogen cycle in general is determined by assimilatory processes in microorganisms and plants, and the use of this primary production by consumers and decomposition by microorganisms (Box **7.4**).

Fig. **7.21** gives a comparison of the annual nitrogen-cycles and the nitrogen levels in various compartments of the ecosystem of a non-tropical prairie and a seasonal tropical savanna (a similar presentation for N-cycles in tropical forests is presented in Fig. **3.26.**).

The compartments distinguished are

- soil, organic N,
- soil, mineralized N,
- roots,
- living epigeous biomass,
- dead litter, and
- atmosphere.

The largest amount of N in either case is in the organic matter of the soil, and it is similar in the prairie and the savanna. The amount of mineral N in the soil is much smaller, and it is somewhat larger in the prairie as compared to the savanna. The amount of N in the roots and in the living epigeous biomass, as well as in the litter, is not very different in the two systems. The rates of N-transfer between the individual compartments, namely absorption of mineral N from the soil, root-shoot translocation, mortality,

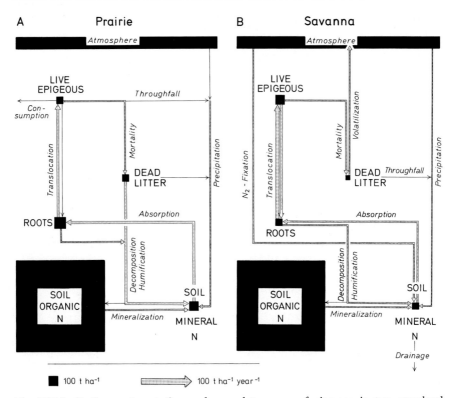

Fig. 7.21A, B. Compartmentation and annual turnover of nitrogen in two grassland ecosystems. **A** *Andropogon gerardi – Andropogon scoparius* prairie in Missouri (USA). **B** Seasonal *Axonopus purpusii-Leptocoryphium lanatum* savanna in Barinas (Venezuela); strongly modified and simplified from Sarmiento (1984; reprinted by permission of Harvard University Press). Sizes of N pools in the different compartments (*boxes*) and transfer rates between the compartments (*arrows*) were drawn to scale to allow direct comparisons of pools and rates both within and between the two ecosystems

decomposition/humification and mineralization, as well as precipitation, throughfall and drainage are similar within the two systems.

It is noticeable that the rate of absorption of mineralized N from the soil is lower in the savanna and that the root/shoot recycling of N (translocation between roots and epigeous biomass) is higher than in the prairie. More important, however, is the presence of two additional transfer processes in the savanna as compared to the prairie, namely

– volatilization and
– atmospheric N_2-fixation.

Volatilization is largely due to fire in the savannas (Sect. **7.4.3**). However, it is also known that tropical soils are significant natural sources of gaseous N-compounds, e.g. in the savannas of the Venezuelan Llanos:

nitrogen oxides	3-13	kg N ha^{-1} a^{-1}
(NO-nitrogen	2.6	kg N ha^{-1} a^{-1})
(N$_2$O-nitrogen	0.65	kg N ha^{-1} a^{-1})
ammonia	11	kg N ha^{-1} during the reproductive period

(Garcia-Méndez et al. 1991; Medina 1993). Atmospheric N$_2$-fixation is an important activity of mats of cyanobacteria between the tussocks of savanna grasses and of free living soil bacteria and root nodule symbioses (Sect. **7.3.1.3**).

In summary, it is surprising how small are the differences between the two grasslands, namely the mesic prairie and the tropical savanna. It should be recalled, however, that diagrams similar to Fig. **7.21** have been drawn for tropical forests (Fig. **3.26**), giving a very different picture. Although the soil organic N is similar, amounts of N in the roots and in the epigeous living biomass are very much larger, and N in the dead litter is somewhat larger in the forests than in the grassland systems. Mineral N in the soil is smaller in the forest. With the exception of precipitation, throughfall and drainage, the rates of N-transfer between the individual compartments are considerably larger in the forests than in the grassland systems. This relates to absorption of mineral N from the soil, root-shoot translocation, mortality, decomposition/humification and mineralization, such that the cycling of N in the forest is much more rapid than in the prairie and the savanna.

7.3.1.2
Nitrogen-Use Efficiency

In Sect. **3.6.2.2** we have already discussed the nitrogen-use-efficiency (NUE) of photosynthesis in relation to the light climate in tropical forests. Again for the savanna grasses, we observe generally linear relationships between levels of N in biomass and rates of photosynthesis (Fig. **7.22**). As mentioned

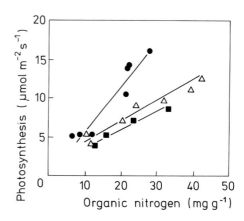

Fig. 7.22.
Relations between rates of photosynthesis and nitrogen content in the biomass of a tropical C$_4$ grass (●, *Panicum maximum*) and two C$_3$ grasses of the temperate zone (Δ *Lolium perenne*; ■ *Festuca arundinacea*). (Medina 1986)

above, there are often differences between species (Sect. **3.6.2.2**). In savannas, the slope of the line for the tropical C_4-grass is much steeper than for the two C_3-grasses of the temperate zone given for comparison (Fig. **7.22**). For crops the ratio of photosynthetic CO_2-fixation to leaf-N also was found to be higher in the C_4-plant maize (1056 μmol CO_2 m^{-2} s^{-1}/mol N) than in the C_3-plant rice (640 μmol CO_2 m^{-2} s^{-1}/mol N). On the other hand, C_4-plants do not necessarily have a competitive advantage over C_3-plants under conditions of low N-supply. Experiments with C_4- and C_3-grasses under natural conditions of a Central European summer have shown that the C_3-grasses tended to be more successful at low N-supply (Gebauer et al. 1987). The authors suggest that this could result from lower transpiration in the C_4-grasses because of the water-saving functions of C_4-photosynthesis (see Sect. **7.2.1**). This would imply lower N input via the transpiration stream from the soil to the shoots. However, some of this effect may be offset by higher temperatures, and the situation may be very different in tropical savannas, and indeed, a higher NUE may be important for C_4-grasses which dominate in the nutrient-poor savannas.

There are even differences between C_4-groups of grasses forming malate and aspartate as the primary CO_2-fixation product. Those species which synthezise the amino-acid aspartate from the oxaloacetate (following CO_2-fixation via PEP-carboxylase) have a higher N-requirement than the malate-forming NADP-malic enzyme group (see Box **7.2**). In S-Africa it was observed that NADP-malic enzyme C_4-plants with their lower N-requirement are characteristic of particularly nutrient-poor, moist savannas while the more N-demanding aspartate-formers (NAD-malic enzyme and PEP-carboxykinase groups) are more frequent in arid savannas.

7.3.1.3
Fixation of Atmospheric Dinitrogen (N₂)

Fixation of N_2 is mediated by the enzyme-complex **nitrogenase** (Box **7.5**). It is restricted to procaryotic microorganisms, bacteria and cyanobacteria ("blue-green algae"), which however, can make important contributions to the N-supply of eucaryotic plants in symbioses and associations.

The symbiotic formation of root nodules with N_2-fixing bacteria (*Rhizobium*) in the Leguminosae is best documented. With N-supply limiting so much the productivity of savannas, one might expect that plants capable of fixing atmospheric dinitrogen (N_2) would be particularly frequent. An important leguminous savanna tree in South America is *Bowdichia* and in Africa various species of *Acacia* (Fig. **7.9**) play an equivalent role. Open woodlands tend to contain more nodulated trees than adjacent forests. There is a progressive increase in the proportion of nodulated trees along a gradient from humid to arid areas, which is negatively correlated with the N-content of the soils (Högberg 1986b). Such a negative correlation is expected. N_2-fixation requires much input of energy, of carbon skeletons for

Box 7.5

Dinitrogen reduction by nitrogenase. (After Kindl 1987)

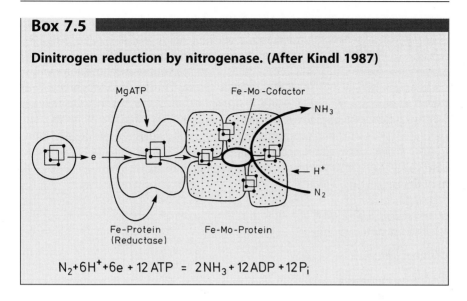

$$N_2 + 6H^+ + 6e + 12\,ATP = 2\,NH_3 + 12\,ADP + 12\,P_i$$

binding of reduced N, and also special morphological differentiation (nodules), and in view of these costs N_2-fixation should not give a competitive advantage when sufficient N is available in the soil. Alternatively, the symbiotic association may be more susceptible to drought stress, although nodulated Leguminosae such as *Prosopis* and *Acacia* are phreatophytes (see Fig. **7.14**).

It has been noted by Ethiopian scientists that in agro-forestry systems it should be sufficient to keep about 40 *Acacia* trees per ha to have sufficient N-fertilization. Although in Africa numerous tree species contribute significantly to the N-budget of to savanna-woodland ecosystems through their N_2-fixation (Högberg 1986b), it is often observed elsewhere that Leguminosae comprise a surprisingly low proportion of biomass in other savanna grasslands. In the Llanos of Venezuela, legumes rarely make up more than 1% of the biomass (Medina 1993). This limitation does not necessarily hold for the number of species of Leguminosae in such savannas. In one case among 127 species 109 were found to be nodulated. The frequency of leguminous species in Venezuela is related to low levels of exchangeable aluminium in the soil (see Sect. **7.3.4.1**) and high levels of exchangeable calcium, i.e. two factors which are inversely related to each other. Overall root-nodule symbioses appear to contribute little to the productivity of this particular savanna system (Medina 1993).

Other factors limit the growth of legume species, and in particular the balance between the supply of nitrogen and other elements must be crucial. Nodulation itself is nutrient limited (Souza Moreira et al. 1992), and the most important limiting nutrient factor in savannas frequently is phosphorus (Högberg 1986a; Sect. **7.3.2**).

The particular demand of phosphorus for root nodules is a well known general phenomenon. In soybean plants the total response of symbiotic N_2-fixation to altered P-supply is a function of both indirect effects on growth of the host plant and more direct effects on the metabolic functions of the nodules (Israel 1993). In Africa, the low availability of phosphorus was found to be a severe restriction for nitrogen-fixing species in moist savannas, and this can explain their low abundance in such ecosystems. The increase in the number of nodulated trees towards drier sites already mentioned above is correlated with a decline in soil-N and an increase in available soil-P so that in African, and perhaps also in Australian savannas, one may distinguish between

– moist/dystrophic and
– arid/eutrophic svannas

(Högberg 1986a).

Phosphorus limitations of nodulation can at least be partially alleviated when the leguminous plants develop a second symbiosis in addition to root nodules, namely mycorrhiza. The fungal hyphae of mycorrhiza enhance nutrient acquisition and may have positive effects, particularly due to increased supply of P to the host plants. In the experiments presented in Table 7.7, nodulation was much more effective when mycorrhiza was present than if production of mycorrhiza was prevented. However, mycorrhiza could be readily replaced by phosphate supply in these experiments. The higher nodulation in plants with mycorrhiza or additional P-nutrition was accompanied by a considerably larger production of total fresh weight and a reduced root/shoot ratio, showing a lower investment in nutrient allocating roots foraging for nutrients and hence larger resource allocation to photosynthezising shoots. In another experiment, where nodulated soybean plants and maize plants were connected by a common mycorrhizal myce-

Table 7.7. Effect of mycorrhiza and additional phosphorus supply on growth and nodulation of legumes (Crush 1974).

Genus		Total FW g	Root/ shoot ratio	Nodules (1-5)	Mycorrhizal infection %	P- and N content (mg/g) P	N
Centro-sema	Mycorrhiza	3.88	0.86	5	86	2.0	28.5
	Non-mycorrhizal	1.67	1.70	1	0	0.5	40.2
	+ phosphate	4.95	0.68	5	0	2.2	30.8
Stylos-anthes	Mycorrhiza	1.63	0.54	5	74	4.4	39.0
	Non-mycorrhizal	0.47	1.12	0	0	2.0	43.1
	+ phosphate	0.91	0.34	5	0	5.8	38.1

lium, it was even observed that the source-sink relations led to P- and N-flows in opposite directions. There was a P-flow from maize to soybean, while soybean could provide maize with N (Bethlenfalvay et al. 1991).

Some tropical tree species have N_2-fixing symbiotic nodules on their leaves and stems. The stem nodules of the tropical leguminous plants *Aeschynomene afraspera* and *Sesbania rostrata* are due to infection by special strains of *Rhizobium*. They are successfully used in green-manuring to improve tropical agriculture in Africa. *S. rostrata* may fix up to 200 kg N ha^{-1} season^{-1} and contribute 50 – 150 kg N ha^{-1} a^{-1} to the soil nitrogen. In this way rice production can be increased two- to threefold (Clarkson et al. 1986).

Besides symbiotic microorganisms, free living bacteria of the genera *Beijerinckia*, *Clostridium* and *Azospirillum* and cyanobacteria ("blue-green algae") may also provide N-input to savanna soils, and possibly in many cases their overall contribution may even be larger than that of root nodules.

Associations of plants with N_2-fixing soil bacteria in the rhizophere may also be of mutualistic benefit, where exudates from the plant roots provide substrates and vitamins and other regulatory compounds to the microorganisms and plants receive N-compounds. In fact, there have been considerable efforts to improve agricultural productivity of tropical grasslands with such associations. Attempts have even been made to use genetic engeneering for the introduction of the nitrogenase-genes (nif$^+$-genes) into some rhizosphere bacteria, which occur more abundantly in the soil than natural N_2-fixing organisms (Hess 1992). The contribution of rhizosphere associations to total N-input may be quite significant in tropical savannas (Table 7.8). Some additional figures for the contribution from bacterial rhizosphere associations are 5 – 18 kg N ha^{-1} a^{-1} in Brasilian grasslands and 78 kg N ha^{-1} a^{-1} in Zimbabwe (Medina 1993).

The contribution of cyanobacteria is shown to be rather modest (Table 7.8). Future research, however, will probably modify this view. In fact, cyanobacteria are extraordinarily abundant in savannas, often forming dense, continuous mats between the tussocks of grasses (Fig. 7.23). The nitroge-

Table 7.8. Nitrogen balances in two humid tropical savannas in South America, Central Venezuela (*Trachypogon* savanna) and in Africa, Ivory Coast. (Medina 1987, 1993)

	Venezuela	Ivory Coast
	(kg N ha^{-1}a^{-1})	
Input through rain	19 (inorganic 4.5)	2.6
Biological fixation		
Blue-green algae	1.4 – 2.5	0.7
Rhizosphere association	9 – 12	6.7
Losses through fire	17 – 23	8.5
Percolation and leaching	5.6	0.5
Balance	+4.9 to + 6.8	+1.0

Fig. 7.23A, B. Mats of cyanobacteria between tussocks of grasses in the Llanos of Venezuela after a longer rainless period (**A** February 1989) and a few days after a rain (**B**
March 1991)

nase is an O_2-sensitive enzyme. In the photosynthezising filamentous cyanobacteria it is located in special cells, the heterocytes, which have thick cell walls limiting O_2 diffusion into these cells, and lack photosystem II and hence photosynthetic O_2-evolution. From the possession of heterocytes, most of the cyanobacteria in the savannas are shown to be N_2-fixers (see also Fig. **8.14**). Indeed, in an example from savannas in Nigeria, where the ground coverage with cyanobacterial mats and crusts was 30%, a much higher value of cyanobacterial N_2 fixation is reported, namely 23 g ha^{-1} d^{-1} in the rainy season, which corresponds to several kilograms per ha over the year (Medina 1993).

A symbiotic system involving cyanobacteria, which has become important in tropical agriculture, is the mutualism between the fern *Azolla* and the cyanobacterium *Anabaena*, which lives in special intercellular air spaces of the fern fronds. *Azolla* grows equally well on freshwater surfaces and on mud and is successfully used for mulching in tropical rice culture.

7.3.2
Phosphorus

Phosphorus has already been mentioned above as one of the most critical nutritional elements in savannas in relation to the P-demand of symbiotic root nodules (Sect. **7.3.1**). Together with N it is so important, because these are the two mineral elements most abundantly and most directly involved in the metabolic machinery of cells. Consequently, P/N ratios have also been used to describe the state of P nutrition. In semi-arid grasslands of the Sahel region of Africa P/N-ratios (mol/mol) of $17 \cdot 10^{-3}$ to $68 \cdot 10^{-3}$ are considered to mark the range within which there is response to P-fertilization, with the lower value characterizing P-deficient and the higher one P-sufficient plants. A more detailed comparison is presented in Fig. **7.24**. Within the "P-responsive" range N-nutrition alone only slightly lowers P/N-ratios, whereas P-nutrition highly stimulates P/N ratios, which are somewhat lower when N + P fertilization was applied.

7.3.3
Biotic Interactions

7.3.3.1
General Overview

Biotic interactions, when considered in relation to the nutrient "stress factor", determine growth, development and productivity of plants in savannas. This includes

– microorganism or fungi-plant interactions,
– plant-plant interactions,
– animal-plant interactions.

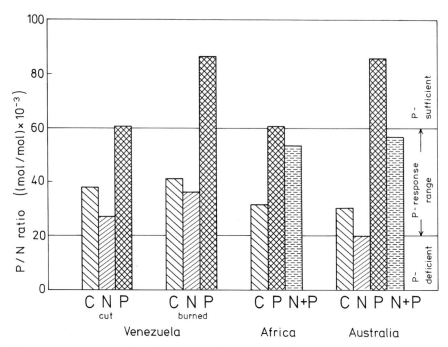

Fig. 7.24. P/N-ratios in the above-ground biomass of different grasslands in the Llanos of Venezuela (cut or burned at the end of the dry season), Africa and Australia (**C** controls), and effects of fertilization with N, P or N + P. (After Medina 1993)

Some of these interactions, mainly representative from the first category, have already been discussed above with respect to N- and P-nutrition, namely leguminous root-nodules, bacterial rhizosphere associations, cyanobacteria dominated cryptogam mats and crusts, and mycorrhiza. Among the nutritional plant-plant interactions the grass-tree relations are of particular interest in savannas. With the extended root systems of trees, tree-biomass may concentrate nutrients from large soil volumes. Due to litter fall and decomposition, the availability of K, Ca and Mg is often higher under tree canopies. The droppings of perching birds may also add to improved nutrient availability in the vicinity of trees (Medina 1993). More sophisticated animal-plant interactions deserve separate sections.

7.3.3.2
Termites and Ants

Mound-building termites excavate and explore large volumes of soil reaching depths of 0.5 to 1 m. In this way they affect soil texture, but in addition they may also enrich nutrients like Ca, K and Mg and to some extent also P. Termites accelerate nutrient recycling and in termite dominated savannas

Fig. 7.25. Leaf-cutter ants

(Sect. **7.1**, Fig. **7.5**) this may make a very considerable contribution to nutrient turnover. In Australia it was observed that termite-mediated mineralization of organic matter may amount to 250 kg ha^{-1} a^{-1} (Medina 1993).

Leaf-cutter ants (Fig. **7.25**) concentrate nutrients in a similar way to mound-building termites. Leaves of grasses and trees are carried into complicated underground chamber systems, where the ants cultivate fungi and play a significant role in nutrient cycling, particularly for deep rooted trees. Observations from Brazil showed that a well developed ant nest can turn over 1 ton of fresh plant weight per year (Medina 1993).

7.3.3.3
Carnivory

Carnivory has already been mentioned in relation to lianas and epiphytes, as a potential strategy for nutrient acquisition (Sect. **4.4.2**). In the temperate climate, carnivorous plants are particularly frequent in moist and acidic sites and especially in peat bogs, which are very poor in nutrients. Similarly, in the tropics, carnivorous plants of the genus *Drosera* are frequently found in great numbers in the wet and often peaty soils of upland herbaceous vegetation types with savanna-like meadows at 1000–2800 m a.s.l. (Fig. **7.26;** see Huber 1988 for site description). The carnivorous genus *Heliamphora* (Sarraceniaceae) is endemic to the Tepuis, the characteristic table mountains of the Guayana highlands in tropical South America.

Fig. 7.26A, B.
Drosera roraimae (A Gran Sabana, Venezuela; B Sierra Maigualida, Venezuela), with droplets of mucilage on the leaf tentacles

Drosera attracts its prey by the numerous brilliant droplets of mucilage ("sun dew") secreted via special glands on the surface of colourful, often reddish, tentacles (Fig. 7.27). The sticky mucilage usually prevents the escape of small insects once they have touched it. The tentacles move in response to mechanical and chemical stimuli caused by the captured animals making thigmotropic and chemotropic as well as thigmo- and chemonastic movements. The prey is thus enveloped and then digested by protea-

Fig. 7.27.
Scheme of a tentacle of *Drosera* after
Gilchrist and Juniper (1974; from Lüttge
1983). The centre of the tentacle is
served by tracheids. It is separated from
the peripheral gland epithelium by an
endodermis, whose radial walls are sub-
erized so that apoplastic transport is
blocked and transport between the
periphery and the interior must use
a symplastic route

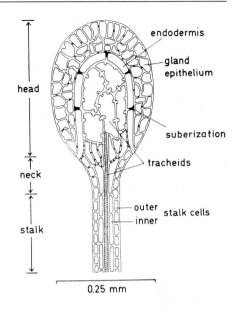

ses secreted from the tentacle glands. Mineral elements like N, S, P, Mg^{2+}, K^+ from the prey then stimulate growth and productivity of the *Drosera* plants (Lüttge 1983).

Heliamphora is a genus of pitcher plants with several species (*H. nutans, H. heterodoxa, H. minor, H. ionasii, H. tatei*). The pitchers are formed of single leaves. They are morphogenetically derived from peltate leaves, so that the interior of the pitcher wall corresponds to the upper leaf surface and the exterior to the lower leaf surface. In the middle of the pitchers there is a small opening, which allows water to flow out and thus prevents over-filling in the high rainfall habitats of *Heliamphora*. Animals are attracted by coloration and nectar secretion at the pitcher orifice. Escape is hindered by hairs and trichomes directed downwards to the bottom. Most of the criteria of true carnivory are fulfilled by all *Heliamphora* species, such as

- attraction of prey through special visual and chemical signals,
- trapping and killing of prey,
- presence of wax scales and other structures preventing escape of prey,
- absorption of nutrients.

Most of the species, however, lack one important trait of true carnivory, i.e. digestive glands and enzyme secretion. In these cases digestion of prey is mediated by bacterial commensals (Schmucker and Linnemann 1959). There is one notable exception though, which is *H. tatei*. In this species there is enzymatic activity in closed pitchers just as they are maturate and open. Since microbes have no access to the closed pitchers, this proves to be

Fig. 7.28A-D

genuine enzyme secretion by the pitcher tissue. Capture of small animals is very effective in *Heliamphora* species in their natural habitat. The carnivorous traits are lost, however, in low light conditions, which indicates that nutrient supply is limiting only under conditions of higher growth rates, and in terms of cost-benefit optimization the sophisticated carnivorous traits are not affordable under limited light (Jaffe et al. 1992). The occurrence of enzyme secretion in only one of the species of *Heliamphora* also suggests evolutionary trends in carnivory within the genus, with enzyme secretion being the most advanced trait in carnivory.

The expression of true carnivory is more dubious in the tanks formed by the leaf rosettes of bromeliads. Jolivet and Vasconcellos-Neto (1993) note that in general, in contrast to dicotyledonous carnivorous plants, among the monocotyledons there is only "protocarnivory" (see Sect. **4.4.2.**). Examples include *Catopsis berteroniana*, *Brocchinia reducta* and *Brocchinia hechtioides* among bromeliads or *Paepalanthus bromelioides* (Eriocaulaceae) of upland plateaus in northern Brazil. In moist upland savannas of Venezuela the terrestrial bromeliad *Brocchinia reducta* shows such extensive developments in some areas, that one may speak of a "*Brocchinia*-savanna". It catches many animals and has a waxy inner surface to prevent escape (Fig. **7.28**). There is breakdown of the bodies of small animals and absorption of solutes via the bromeliad scales. The outer walls of the scale cells have an unusual structure. They have a labyrinthine-like appearance and particularly large pores (6.6 nm) allowing the passage of rather large molecules, which possibly is followed by cellular uptake via endocytosis-vesicles (Owen and Thomson 1991). The species has been considered as a true carnivorous plant (Givnish et al. 1984), although glands and enzyme secretion are totally absent.

7.3.4
The Aluminium Problem

7.3.4.1
The Aluminium Load in Tropical Soils

Clay minerals typical for savanna soils are

- kaolinite, $Al_2O_3 \cdot SiO_2 \cdot 2H_2O$;
- gibbsite, $Al_2O_3 \cdot 3H_2O$;
- goethite, $FeO(OH)$.

◄**Fig. 7.28A-D.** *Brocchinia reducta* in a wet marshy savanna (**A**), with the typical bromeliad inflorescence (**B**), with the wax on the adaxial leaf surface that should prevent the escape of animals fallen into the bromeliad tank (**C**), and a tank cut open to show the putrefying mass of animals at the bottom (**D**). (Gran Sabana, Venezuela; February 1989)

They have low cation-exchange capacity (CEC) and low water storage capacity. Ferralization is a frequent process where bases and silicious acid are leached, leaving aluminium and iron oxides (Al_2O_3, Fe_2O_3). Thus, ferralitic soils always have very high concentrations of Al^{3+}. A special formation is the "arecife" in the Llanos of Venezuela (see above Sect. 7.2.2 and Fig. 7.14). Iron oxide is precipitated at high groundwater table level in the young alluvial sediments forming these soils, such that gravel, sand and clay are solidified to a hard crust of a thickness of 1–3 m. It generally lies at a debth of 30–80 cm but may also be lower or closer to the surface (Fig. 7.14). Soils of the Brazilian cerrados contain between 75 and 360 ppm Al^{3+} (Eiten 1972).

By comparison, high Al-load in acidifying soils has also been observed in the temperate zone. It is thought to be one of the possible reasons for forest decline, and here, the equilibrium soil solution contains up to 20–40 ppm Al.

7.3.4.2
Potential Damage to Plants by Aluminium

Damage of plants by high aluminium levels in the substratum may be largely due to extracellular effects on the surface of roots and cells, although uptake and translocation of aluminium in the plants may also be involved. The major interactions known so far are the following (Lüttge and Clarkson 1992):

- **Ionic interactions:**
  - **Phosphate. Al precipitates phosphate at surfaces in the apoplast in the form of the hardly soluble $Al_2 (PO_4)_3$ salt and thus reduces P-availability.**
  - **Iron.** High Al levels in the medium are associated with high acidity, which at the same time leads to increased mobility of iron and thus may induce Fe-stress.
  - **Divalent cations,** Ca^{2+}, Mg^{2+}, Mn^{2+}, Zn^{2+}. Al occupies important cation-exchange sites in the apoplast and thus prevents access of essential divalent cations to these sites, which adversely reduces their availability to the plants.

- **Membrane interactions:**
  - Al may bind to membranes and affect their **structure and fluidity**; by increasing the rigidity of membranes Al may change permeability.
  - Al may block **K^+-channels** in membranes.
  - Al may inhibit **proton-pumping ATPases** of membranes.

- **Interactions with intracellular messenger networks:**
  - Al may become involved in Ca^{2+}/calmodulin interactions, which are important in intracellular regulatory processes and signalling at the molecular level.

In conclusion, overall the effects of aluminium are a disturbance of mineral nutrition of plants across a range of levels.

7.3.4.3
Plant Responses

Aluminium has different consequences for tropical forests as compared to savannas. In the forests, even when Al-concentrations in the soil are high the effect on plants is smaller, because the nutrient cycle is tightly coupled between the decomposing litter and the vegetation and tends not to involve the mineral soil very much. In savannas, however, plants take up minerals from the soil solution, which is in equilibrium with an Al-enriched exchange complex. Since most forest-tree species are more sensitive to aluminium than savanna plants, the Al-load of soils can in part explain the competition between forest, cerrados and savannas (Eiten 1972). While the phenomenon of a high Al-load in tropical soils has long been documented, it may be one important determinant of the complex ecological regulation leading to the co-occurrence of forest and savanna in the tropics (Medina 1982).

Among trees in a cloud forest of Northern Venezuela there are Al-accumulators and Al-excluders, reflected in the xylem sap concentration in Al. In the Al-accumulator *Richeria grandis* (Euphorbiaceae), Al-levels in the leaves increased with age to levels of about 15 000 ppm (Cuenca et al. 1990, 1991). The gallery-forest tree *Vochysia venezolensis* (Vochysiaceae) in South America also accumulates up to 25 000 ppm Al related to dry matter (Eiten 1972; Sarmiento 1984). In a savanna in Trinidad, Al-levels in the grass *Panicum stenoides* were on average 910 ppm with maximum levels over 4 000 ppm, and the herbaceous Melastomataceae *Acisanthera uniflora* contained over 20 000 ppm (Sarmiento 1984). Haridasan (1982) lists Al-levels between 4 000 and 14 000 ppm for various Al-accumulating cerrado species of central Brazil, but the highest levels of Al in plants quoted from the literature are 66 100 ppm for the Melastomataceae *Miconia acinodendron* and 72 300 ppm for the Symplocaceae *Symplocos spicata*. For comparison, in areas of forest decline in the temperate zone, Al-levels in the root dry mass range from 20 to 14 000 ppm depending on sites and soil depths (Lüttge and Clarkson 1992).

The potential effects of aluminium toxicity, as listed above (Sect. **7.3.4.2**), may not be always realized individually or in combination. They may be modulated by other edaphic factors and also responses vary between plant species and within individuals. In many cases Al-accumulation is not necessarily correlated with reduced uptake of other cations (Haridasan 1982), and Al-concentrations which severely inhibited root growth did not affect Ca^{2+} uptake, suggesting that rhizotoxicity is a function solely of Al^{3+}-activity at the root-cell membrane surface (Kinraide et al. 1994; Ryan et al. 1994). For *Miconia albicans* (Melastomataceae) from the cerrados of Brazil, conditions which lead to a degree of aluminium accumulation, such as non-calcareous acid soils, are even favourable for growth (Haridasan 1988).

Clearly there are distinctions between plants with Al inclusion and exclusion, respectively, which are related to taxonomic groups. However, both may have mechanisms to alkalinize the rhizosphere, which diminishes Al-mobility and avoids the stress. The "includers" may also have various mechanisms for tolerating high internal Al-levels, such as Al-chelation and desposition in the vacuoles. Thus, both avoidance and tolerance help to sustain plant life at high ambient Al-levels in savannas.

7.4
The Fire Factor

7.4.1
The Causes of Fire: Anthropogenic and Natural

Fires play an important role in tropical biota (Goldammer 1990; Fig. **7.29**). Currently the major cause of fires is man. Alexander von Humboldt (1808/1982) recognized this in his "Journey to South America" and mentions benefits and even the pleasure of fires, but also suggests the drawbacks.

"The pastoral people burn the grassland to obtain fresher and finer grass by new growth ... Thus, if one relaxes on a magnificent tropical evening at the shore of the lake[5] and enjoys the delightful coolness, one observes with pleasure the picture of the fires along

Fig. 7.29. Man-made fires from Martius' Flora Brasiliensis (1840–1906)

---

[5] Lake of Valencia, Venezuela.

the horizon, reflected in the waves beating the shore. ... The savanna is frequently burnt to improve the pasture ever since the Llanos were inhabited. Together with the grasses by chance the scattered groups of trees are also destroyed. No doubt, these plains in the 15th century were not as bare as now. Nevertheless, even the first conquerers coming from Coro describe the savannas, in which one sees nothing but sky and grass, widely tree-less and difficult to pass because of the heat reflected by the ground."[6]

Fires ignited by man have not only been used in slash-and-burn agriculture (Sects. **1.3** and **3.3.1.3**) or in the management of pastures (Sect. **7.4.3**) but also by very early hunter/gatherer societies to drive game out of forest thickets. However, fires in savannas, as well as in other tropical and non-tropical ecosystems, may also be caused naturally. In certain dry savannas of Africa during storms there is often lightning with little rain or before the rain sets in. In Australia, fire has been long considered as a natural environmental stress factor (Walter and Breckle 1984). Palaeontological findings show that fires destroying vegetation must have occurred since the Devonian (376×10^6 years ago). The prerequisites for ignition of such fires are

- a certain minimal atmospheric concentration of oxygen and
- the presence of combustible material.

The minimal O_2-concentration required was shown experimentally to be 13 %, i.e. slightly less than 2/3 of the present level. Since it is assumed that atmospheric oxygen has resulted from photosynthetic O_2-evolution, we may conclude that 13 % must have been reached by the time of the Devonian. Terrestrial vegetation had also developed to a stage that the second of the above criteria was fullfilled. As the possible causes we may list

- lightning,
- sparks formed during rockfalls,
- vulcanism, and
- self-ignition of fermenting material

(Walter and Breckle 1984; Jones and Chaloner 1991). Thus, plants have been exposed to fire for long enough to allow evolution of special adaptations with stress avoidance and resistance in an ecophysiologically defined group of plants called **pyrophytes**.

---

[6] Südamerikanische Reise
"... brennt das Landvolk die Weiden ab, um ein frischeres, feineres Gras als Nachwuchs zu bekommen... Wenn man so an einem herrlichen tropischen Abend am Seeufer[5] ausruht und die angenehme Kühle genießt, betrachtet man mit Lust in den Wellen, die an das Gestade schlagen, das Bild des roten Feuerrings am Horizont... Seit die Llanos bewohnt... sind, zündet man häufig die Savanne an, um die Weide zu verbessern. Mit den Gräsern werden dabei zufällig auch die zerstreuten Baumgruppen zerstört. Die Ebenen waren ohne Zweifel im 15. Jahrhundert nicht so kahl wie gegenwärtig; indessen schon die ersten Eroberer, die von Coro herkamen, beschrieben die Savannen, in denen man nichts sieht als Himmel und Rasen, im allgemeinen baumlos und beschwerlich zu durchziehen wegen der Wärmestrahlung des Bodens."

7.4.2
Pyrophytes

Discussing adaptations of plants to fire as an important natural ecological factor it is useful to distinguish between

- **pyrophilous plants,** which obtain an advantage in the competition with other plants, and
- true **pyrophytes,** which essentially need fire at least at some stage in their life cycle.

Smaller plants may survive fires since the temperature at the soil surface may reach values of ~75 °C only for a few minutes, and 1 to 5 cm below the surface temperatures may already be much lower (Walter and Breckle 1984). Thus the terminal buds in the center of tussock grasses are well enough protected, and regeneration can also occur from below-ground plant organs. Taller fire resistant plants, apart from specialized savanna trees with thick bark and dormant buds (see Sect. **7.4.3**), are often tree ferns or monocotyledonous plants (like palms, *Yucca* or *Xanthorrhoea*, Fig. **7.30**). These plants do not have a cambium at the periphery of their stems, as found in

Fig. 7.30A, B. Pyrophilous plants of Australia. **A** *Xanthorrhoea* (Liliaceae), **B** *Cycas media* (Cycadaceae)

dicotyledonous shrubs and trees. In some *Eucalyptus* species in Australia, survival is guaranteed by formation of below-ground stem-thickenings ("lignotubers"), and reproduction by seeds is facilitated by removal of dry litter during fires and by killing the predators of young seedlings (Walter and Breckle 1984).

The genuine pyrophytes are literally dependent on fire. A member of the Australian Liliaceae (or Xanthorrhoeaceae), *Xanthorrhoea,* only flowers after a fire. Among the woody plants of the cerrados in Brazil, Coutinho (1976) distinguished the following responses of flowering to fires:

- species which quantitatively and qualitatively depend on fire and where fire elicits flowering at any time during the seasons;
- species where fire elicits flowering only during the dry season or in combination with short days;
- species which do not react to fire and flower during the dry season or after induction by short days;
- species which are damaged by fire and normally flower during the rainy season or after induction by long days.

Among the Australian Proteaceae there are many species where the fruits can only open and disperse seeds after a fire, e.g. *Banksia ornata, Hakea platysperma* and *Xylomelum pyriforme,* as well as the conifer *Actinostrobus.*

The evolution of such fire resistant and fire demanding plants also implies that fire is necessary to stabilize the ecological equilibria in ecosystems which have always been regularly subject to fire. In fact, it has been noted in some conservation areas and national parks that total prevention of fires has had adverse effects (Walter and Breckle 1984).

7.4.3
Burning by Man: Losses and Gains

During severe drought periods in savannas the decomposition of above-ground dead organic matter by microorganisms is very much reduced. This cover prevents new growth. Perennial grasses die back and seedling mortality under such a dense layer of dead plant material is high. Eventually the whole grass layer may die, as shown in a long-term experiment over 20 years, where specific areas were protected from fire and grazing at the biological field station at Calabozo in the Venezuelan Llanos (Medina and Silva 1990).

Rapid mineralization by fire removes the dead biomass and also has nutritional effects. Burning decreases soil acidity, and promotes mineralization of nitrogen. After an episode of fire, rates of nitrification increase for several years, followed by a decline of nitrification and increase in ammonium availability (Stewart et al. 1993). However, fire not only enriches the soil with minerals, it may also lead to losses especially of N and S in the

Table 7.9. N- and S-input and losses and biomass production in a *Trachypogon* savanna in central Venezuela. (Medina 1982)

Biomass production	Losses as volatile gases		Input by rain	
	N	S	N	S
(t ha^{-1})	(kg ha^{-1})		(kg ha^{-1})	
Minimum 2.9	9.3	2.6	1.3	6.8
Maximum 10.0	32.0	9.0	4.7	9.1

form of volatile oxides (Table **7.9**). Most of these losses are from vegetation rather than soil (Stewart et al. 1993). For N the range of such losses is

- 4.5–5.6   kg ha^{-1}a^{-1} in Australia,
- 8–10   kg ha^{-1}a^{-1} in Africa, and
- 8   kg ha^{-1}a^{-1} in Venezuela

(Medina 1993). It may much exceed the import via rain (Tables **7.8** and **7.9**).

The global role of fires has been surveyed by Fontan (1993). Table **7.10** shows the annual turnover of biomass in forests and savannas in the tropical regions of the world. The contribution of savanna fires to the total biomass burnt per year is seen to be high in America but particularly so in Africa. Fires not only cause losses of minerals but also make a significant contribution to atmospheric loading of infrared-active gases which cause the green-house effect (CO_2, CO, CH_4, O_3) and with straighforward pollutants (like CO, N-oxides, O_3) (Table **7.11**).

Frequent man-made fires also open the soil surface to solar radiation, which leads to oxidation and burning of humus (Eiten 1972) and leaching following rainfall. Therefore **in dry savannas fire is always detrimental**.

Moreover, of course, fire always damages the **forests** unless it is wet gallery forest with permanently inundated soil (Fig. **7.31**). The fires intrude from the edges into the forests, and where the trees are not fire-resistant,

Table 7.10. Burnt biomass in the tropical regions of the world from tropical forests, savannas and other sources (firewood plus agricultural waste) in Tg dry matter per year. (After Fontan 1993)

Tropical region	Forests	Savannas	Other sources
America	590 (34 %)	770 (44 %)	370 (22 %)
Africa	390 (12 %)	2430 (76 %)	400 (12 %)
Asia	280 (13 %)	70 (3 %)	1840 (84 %)
Oceania	– (0 %)	420 (94 %)	25 (6 %)
Total	1260 (17 %)	3690 (49 %)	1367 (34 %)

Table 7.11. Global gas emissions form bushfires in Tg a^{-1} and in % of total global emissions. (Fontan 1993)

Compound	Emissions from bush fires	
Carbon dioxide (CO_2)	3500	40 %
Carbon monoxide (CO)	350	32 %
Methane (CH_4)	38	10 %
Hydrocarbons other than methane	24	24 %
N-oxides	9.3	27 %
Ammonia (NH_3)	5.3	12 %
Chloromethane (CH_3Cl)	0.5	22 %
Hydrogen gas (H_2)	19	25 %
Ozone (O_3)	420	38 %

year by year savanna gradually encoraches into the area previously occupied by forests (Fig. **7.32**).

Savanna **trees** are particularly fire resistant (Figs. **7.33** and **7.34**), being predominantly evergreen, with a thick corky bark and dormant buds, and sprout after fires before the beginning of the rainy season (see above Sect. **7.2.2**). Some trees, in areas where regular burning occurs, may even restrict formation of their woody stems to below the ground surface, where dor-

Fig. 7.31. Savanna in the Llanos of Venezuela near Puerto Ayacucho with scattered islands of semi-evergreen forest and wet gallery forest (background to the right)

Fig. 7.32. Remains of montane forest near Akanzobe in Madagascar. The picture of B in the centre shows a zone of common brake fern (*Pteridium aquilinum*) between the grassland and the forest, which burns very readily and gradually progresses towards the forest. (Photographs courtesy M. Kluge)

Fig. 7.33.
Palicourea rigida in the
Llanos of Venezuela with a
thick corky bark coloured
black from fire (January
1989)

mant buds are protected and can readily produce new growth after a fire
(Fig. **7.34B**). Deciduous trees, with phenological cycles related to the seasonality of rainfall, are more sensitive to fire, and they are excluded from
regularly burned savannas (Medina and Silva 1990). In the experiment at
Calabozo mentioned above, during 20 years of protection, total tree density
increased considerably, both of fire-resistant savanna trees and fire-
sensitive species from the surrounding semideciduous forest (Table **7.12**),
and Table **7.13** shows similar findings for a Brazilian cerrado.

   In wet savannas fire can be beneficial, but only when the timing is correct. If a fire occurs before the start of the rainy season, the trees are protected (see above) and safe sprouting of grasses is obtained. The centers of
tussocks of grasses supporting the meristems for regrowth are protected
from the heat of the fire by an insulating layer of old leaves. If **burning**
occurs **too early in the dry season,** subsequent new growth uses up water
reserves and dies before the rainy season sets in, and the whole plant may
dry out totally. If **burning** is done **too late in the dry season,** new growth is
induced when the water reserves are already exhausted and growth is very

Fig. 7.34A, B. *Byrsonima verbascifolia.* **A** in the cerrados of Brazil with bark coloured black from fire (August 1993); **B** in the Llanos of Venezuela with fresh leaves and flowers sprouting from an underground stem (February 1989)

Table 7.12. Number of tree stems per ha in a fire-protected savanna plot over 21 years at Calabozo, Venezuela. (Medina and Silva, 1990)

Species	1962	1969	1977	1983
Savanna trees (evergreen)	92	174	270	1010
Forest species (evergreen)	1	3	11	32
Forest species (deciduous)	0	77	229	1319
	93	254	510	2361

Table 7.13. Comparison of adjacent hectares of low-tree and scrub cerrado of central Brazil periodically burned each 2-3 years and not burned for over 20 years, respectively. (Density of stems is larger than density of individuals since some individuals produce more than one stem), (G. Eiten and R. H. R. Sambuichi, pers. comm.)

		Not burned	*Burned*
Stems	(number ha^{-1})	6677	1765
Individuals	(number ha^{-1})	5788	1663
Species		92	57

limited. If **burning** occurs **in the middle of the dry season**, green biomass is produced, which is maintained until the beginning of the rainy season (Medina 1982; Medina and Silva 1990). The experiments at Calabozo have shown that maximum above-ground biomass in a protected savanna increased during 4 years after the last fire (Table **7.14**) and then stabilized at a certain level (Fig. **7.35**). Fire given later at the beginning of the rainy season, led to lower biomass production than fire before the middle of the dry season (Table **7.15**). The seasonal development of the grass *Trachypogon plumosus* shows that the green biomass after a fire is somewhat increased as

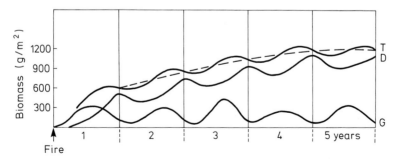

Fig. 7.35. Variations in the epigeous biomass (*T* total; *G* green; *D* dry) of a savanna at Calabozo, Venezuela, after it was burned and until it reached a steady state (*dotted line*) in 5 years. (Sarmiento 1984; reprinted by permission of Harvard University Press)

Time since last fire (y)	Maximum epigeous biomass (g m^{-2})
1	230–730
2	520–850
3	980
4	1200
5	1200

Table 7.14.
Maximum epigeous biomass in a fire-protected *Trachypogon* savanna at Calabozo, Venezuela. (Sarmiento 1984)

Table 7.15. The effect of fire given at different times during the year on daily biomass production in a savanna at Calabozo, Venezuela. (Medina 1982)

	Biomass production (g m^{-2} d^{-1})
3-4 years protected from fire	2.5–2.6
Fire before the middle of the dry period	2.9–3.7
Fire at the beginning of the wet period	1.8–2.1

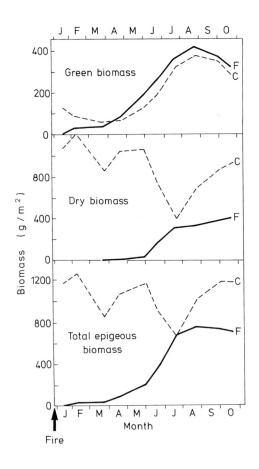

Fig. 7.36.
Variations in the epigeous biomass of a savanna at Calabozo, Venezuela, during the first year after a fire (*F*) and in a non-burnt control plot (*C*). (Sarmiento 1984; reprinted by permission of Harvard University Press)

compared with the control (Fig. **7.36**) and new dry biomass and total above-ground biomass increase rapidly over the year after most of the old biomass was destroyed by the fire.

7.5
Productivity and Nutrient Categories of Savannas

To conclude the major part of this Chap., we should consider overall productivity and nutrient categories of savannas. Table **7.16** gives some data for productivity of seasonal and hyperseasonal savannas. Conditions required for intensive and productive agriculture in the tropics may be available as listed after Eiten (1972):

i)   **above rocks,** which lead to soils rich in minerals, e.g. limestone and vulcanic rocks;
ii) **in flood-plains** with periodically rising water-tables, leading to repeated renewal of the mineral content of the soils, e.g. in the Amazonas, the Nile, the Ganges;
iii) strong **organic fertilization;**
iv) use of **chemical fertilizers.**

Limitations to the third and fourth possibilities are the economic costs, and with only small areas available for organic fertilization. Nutrient categories of savannas depend on the nutrient levels in the soil and the recirculation rates. Fertility of savannas may be estimated and ranked by the sum of extractable bases

$$\Sigma\ (K^+ + Ca^{2+} + Mg^{2+} + Na^+)\ [cmol(+)\ /\ kg(soil)],$$

Table 7.16. Productivity of savannas, biomass in g m^{-2}. (Sarmiento 1984)

Savannas	Above ground	Below ground	Total
Seasonal	800 – 1300	600 – 800	1400 – 2100
Hyperseasonal	800 – 1400	900 – 1100	1700 – 2500

Table 7.17. Nutrient categories of savannas. (After Sarmiento 1984)

		← Amount in the soil →	
Rate of recirculation		Large	Small
	Slow	Ca, Mg, Na	Various elements
	Rapid	K can be limiting	P most strongly limiting

where values lower 5 cmol (+) kg^{-1} mark dystrophic and values higher than 20 cmol (+) kg^{-1} eutrophic savannas, with mesotrophic savannas in between (Medina 1993). Generally, however, phosphorus is often the most strongly limiting element, although even potassium can become limiting in situations of rapid recirculation (Table 7.17). For nitrogen levels in soil and vegetation and the turnover rates, comparisons between savannas and forests have already been made above (Sect. 7.3.1.1, Figs. 3.28 and 7.21). Table 7.18 provides another more detailed comparison, including more elements. It shows, in this case, the much poorer state of the savanna as

Table 7.18. Comparison of the nutrients in a forest and a savanna soil in Nicaragua, both profiles on the same piedmont deposit. (Data from Alexander 1973)

Soil horizon	Organic C (%)	Total N (%)	Cation exchange capacity (meq/100 g)	Ca	Mg (meq/100 g)	K
			Rainforest			
A11	5.3	0.53	36.7	4.29	3.53	0.68
A12	4.1	0.39	26.2	1.90	1.91	0.43
A3	2.7	0.26	19.1	0.26	0.79	0.15
B1	0.9	0.12	11.0	Trace	0.56	0.16
B21	0.5	0.08	14.0	0.05	1.07	0.07
B22	0.2	0.07	12.3	0.11	0.89	0.05
B23	0.2	0.06	17.9	0.11	1.06	0.05
			Savanna			
A1	2.1	0.14	10.3	Trace	0.29	0.05
A2	1.0	0.07	6.2	Trace	0.23	0.04
B21	0.9	0.08	8.1	Trace	0.25	0.04
B22	0.5	0.05	9.3	Trace	0.29	0.03
B23	0.3	0.02	8.4	Trace	0.25	0.03
B24	0.1	0.01	8.1	Trace	0.29	0.04

	Relation of vegetation versus soil
Forest:	
K	>
P	>
N	≅
Ca, Mg	≷
Savanna:	
K	<<
P	=
N	<<
Ca, Mg	<<

Table 7.19.
Nutrient distribution between vegetation and soil in a tropical forest and a savanna. (After Sarmiento 1984)

compared with the forest. The typical distribution of some mineral elements between the vegetation and the soil in tropical forests and savannas is schematically summarized in Table **7.19**. In the savanna most of the K, N, Ca and Mg is in the soil, whereas in the forest most K and P is contained in the vegetation. For forests N is about equally distributed between vegetation and soil, similarly to P in savannas.

7.6
Savanna-Desert Interfaces:
the Sahel Problem as a Case Story

We have, in this Chap., often talked about interactions between savannas and forests. To the drier side, on the other hand, there is another important interface, namely that between savanna and desert. The large deserts of the world lie mainly outside the tropics,

18°30′ North
(West-Sahel)

15°41′ North
(East-Sahel)

14°52′ North
(West-Sahel)

13°34′ North
(East-Sahel)

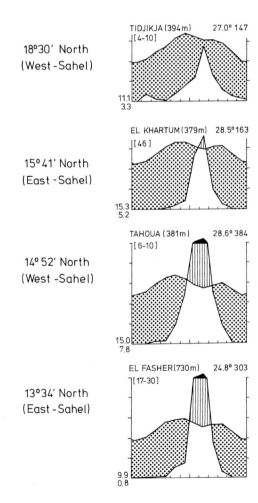

Fig. 7.37.
Klimadiagramm graphs of four stations in the Sahel zone. (Walter and Breckle 1984, with kind permission of S.-W. Breckle and G. Fischer-Verlag)

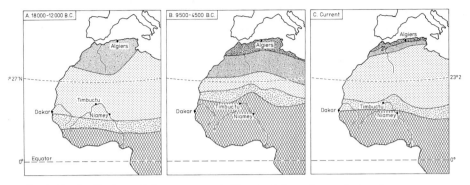

Fig. 7.38. Oscillations of the area occupied by the Sahara desert during the past 18 000 years. (Petit-Maire 1984, with kind permission of La Recherche)

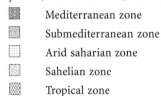

Mediterranean zone

Submediterranean zone

Arid saharian zone

Sahelian zone

Tropical zone

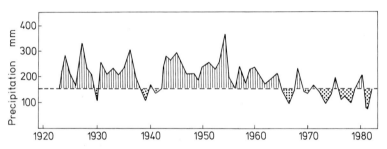

Fig. 7.39. Annual means of precipitation in the Sahel zone from 1922 to 1984 related to the conventional Saharo-Sahelian transition line given here at 150 mm (*dashed horizontal line*); wetter years ⃞ , drier years ⃞ . (Petit-Maire 1984, with kind permission of La Recherche)

and thus, it is not within the scope of this book to treat the ecophysiology of desert plants. It is interesting, however, to consider briefly the tension zone (ecotone) between savanna and desert in addition to that between savanna and forest.

To do this, desertification in the Sahel region of Africa offers itself as an appropriate case story as it has caused much public concern due to the dramatic economic and social implications. The arab word "sahel" in fact means coast or shore, referring to the southern delineation of the sand-ocean of the Sahara, a pertinent way to portray the ecotone. The region is characterized by summer rain with an 8–10 months long drought period. According to the annual precipitation a north-south zonation is given as follows

- saharo-sahelian transition zone,   100–200 mm;
- sahelian zone,   200–400 mm;
- sudano-sahelian transition zone,   400–600 mm

(Fig. **7.37**, Walter and Breckle 1984). The latitudinal position of the saharo-sahelian transition zone, the savanna-desert ecotone, may vary due to climatic oscillations. The case of the sahel is particularly interesting because very strong climatic oscillations have been documented both for the extended period of the last 30 000 years and for much shorter intervals in our century.

Fig. **7.38** illustrates the large changes of the area occupied by the Sahara over the ages. During the last ice age (18,000–12,000 y B.C.) the Sahara had enlarged considerably and then, in the post-glacial period (9,500–4,500 y B.C) contracted again. At that time the river Niger near Timbuctu had a large inland delta with a flooding plain of 20,000 km^2. At present the Sahara again occupies a large area similar to that in the last ice age (Petit-Maire 1984).

The stochastic appearance of drought periods interchanging with wetter intervals or the movement of the saharo-sahelian transition zone more to the south and more to the north respectively, during our century is shown in Fig. **7.39**. Remote-sensing of the vegetation density (Sect. **2.3**) resolves differences for individual years. The long wet period between 1942–1966 led to the extension of savannas and was followed by an increase of the human pupulation and the herds of the nomads. However, this prepared the ground for the catastrophes of the years after 1966. During the increasingly frequent and extended drought periods the land could not sustain the population growth any longer. The example is both tragic and illustrative. It shows that it is impossible to make long term prognoses on the basis of few singular events during periods of short or medium duration.

References

Alexander EB (1973) A comparison of forest and savanna soils in north-eastern Nicaragua. Turrialba 23:181–191

Baruch Z, Fernández DS (1993) Water relations of native and introduced C$_4$ grasses in a neotropical savanna. Oecologia 96:179–185

Bethlenfalvay GJ, Reyes-Solis MG, Camel SB, Ferrera-Cerrato R (1991) Nutrient transfer between the root zones of soybean and maize plants connected by a common mycorrhizal mycelium. Physiol Plant 82:423–432

Black CC (1973) Photosynthetic carbon fixation in relation to net CO$_2$ uptake. Annual Rev Plant Physiol 24:253–286

Clarkson DT, Kuiper PJC, Lüttge U (1986) Mineral nutrition:sources of nutrients for land plants from outside the pedosphere. Prog Bot 48:80–96

Coutinho LM (1976) Contribuiçao ao conheciamento do papel ecologico das queimadas na floraçao do especias do cerrado. Tesc de livre-docente en Ecologia Vegetal. Universidade de Sao Paulo, Sao Paulo

Crush JR (1974) Plant growth response to vesicular-arbuscular mycorrhiza. VII. Growth and nodulation of some herbage legumes. New Phytol 73:743–752

Cuenca G, Herrera R, Medina E (1990) Aluminium tolerance in trees of a tropical cloud forest. Plant Soil 125:169–175

Cuenca G, Herrera R, Mérida T (1991) Distribution of aluminium in accumulator plants by x-ray microanalysis in *Richeria grandis* Vahl leaves from a cloud forest in Venezuela. Plant Cell Environ 14:437–441

Eiten G (1972) The cerrado vegetation of Brazil. Bot Rev 38:201–341

Eiten G (1986) The use of the term "savanna." Trop Ecol 27:10–23

Fontan J (1993) La pollution atmosphérique sous les tropiques. Recherche 24:400–408

Garcia-Méndez G, Maass JM, Matson PA, Vitousek PP (1991) Nitrogen transformations and nitrous oxide flux in a tropical deciduous forest in Mexico. Oecologia 88:362–366

Gebauer G, Schubert B, Schumacher MI, Rehder H, Ziegler H (1987) Biomass production and nitrogen content of C$_3$- and C$_4$-grasses in pure and mixed culture with different nitrogen supply. Oecologia 71:613–617

Gilchrist AJ, Juniper BE (1974) An excitable membrane in the stalked glands of *Drosera capensis* L. Planta 119:143–147

Givnish TJ, Burkhardt EL, Happel RE, Weintraub JD (1984) Carnivory in the bromeliad *Brocchinia reducta*, with a cost/benefit model for the general restriction of carnivorous plants to sunny, moist, nutrient-poor habitats. Am Nat 124:479–497

Goldammer JG (ed) (1990) Fire in the tropical biota. Ecosystem processes and global challenges. Ecological studies vol 41. Springer, Berlin Heidelberg New York

Haridasan M (1982) Aluminium accumulation by some cerrado native species of central Brazil. Plant Soil 65:265–273

Haridasan M (1988) Performance of *Miconia albicans* (Sw.) Triana, an aluminium-accumulating species, in acidic and calcareous soils. Commun Soil Sci Plant Anal 19:1091–1103

Hatch MD (1987) C_4 photosynthesis:a unique blend of modified biochemistry, anatomy and ultrastructure. Biochim et Biophys Acta 895:81–106

Hatch MD, Osmond CB (1976) Compartmentation and transport in C_4 photosynthesis. In: Stocking CR, Heber U (eds) Transport in plants III. Intracellular interactions and transport processes. Encyclopedia of plant physiology NS. vol 3. Springer, Berlin Heidelberg New York 1976, pp 144–184

Hess D (1992) Biotechnologie der Pflanzen. Ulmer, Stuttgart

Högberg P (1986a) Soil nutrient availability, root symbioses and tree species composition in tropical Africa: a review. J Trop Ecol 2:359–372

Högberg P (1986b) Nitrogen-fixation and nutrient relations in savanna woodland trees (Tanzania). J Appl Ecol 23:675–688

Huber O (1982) Significance of savanna vegetation in the Amazon territory of Venezuela. In: Prance GT (ed) Biological diversification in the tropics. Proc Vth Int Symp Assoc Trop Biol, Venezuela 1979, Columbia Press, New York, pp 221–244

Huber O (1987) Neotropical savannas:their flora and vegetation. Trends Ecol Evol 2:67–71

Huber O (1988) Guayana highlands versus Guayana lowlands, a reappraisal. Taxon 37:595–614

Huber O (1990) Savannas and related vegetation types of the Guayana shield region in Venezuela. In: Sarmiento G (ed) Las sabanas americanas, aspecto de su biogeografia, ecologia y utilizacion. Centro de Investigaciones Ecológicas de los Andes Tropicales, Universidad de Los Andes, Mérida, Venezuela, pp 57–97

Israel DW (1993) Symbiotic dinitrogen fixation and host-plant growth during development of and recovery from phosphorus deficiency. Physiol Plant 88:294–300

Jaffe K, Michelangeli F, Gonzalez JM, Miras B, Ruiz MC (1992) Carnivory in pitcher plants of the genus *Heliamphora* (Sarraceniaceae). New Phytol 122:733–744

Jolivet P, Vasconcellos-Neto J (1993) Convergence chez les plantes carnivores. Recherche 24:456–458

Jones TP, Chaloner WG (1991) Les feux du passé. Recherche 22:1148–1156

Kindl H (1987) Biochemie der Pflanzen. Springer, Berlin Heidelberg New York

Kinraide TB, Ryan PR, Kochian LV (1994) $Al^{3+}-Ca^{2+}$ interactions in aluminium rhizotoxicity. II. Evaluating the Ca^{2+} -displacement hypothesis. Planta 192:104–109

Lüttge U (1983) Ecophysiology of carnivorous plants. In: Lange OL, Nobel PS, Osmond CB, Ziegler H (eds) Physiological plant ecology. III. Responses to the chemical and biological environment. Encyclopedia of plant physiology NS. vol 12 C. Springer, Berlin Heidelberg New York, pp 489–517

Lüttge U, Clarkson DT (1992) Mineral nutrition: aluminium. Prog Bot 53:63-77

Lüttge U, Kluge M, Bauer G (1994) Botanik, 2. Aufl. VCH, Weinheim

Mariotti A, Peterschmitt E (1994) Forest savanna ecotone dynamics in India as revealed by carbon isotope ratios of soil organic matter. Oecologia 97:475–480

Martius CFP (1840-1906) Flora brasiliensis, vol 1–15. München and Leipzig

Medina E (1982) Physiological ecology of neotropical Savanna plants. In: Huntles BJ,

Walker BH (eds) Ecological studies, vol 42: Ecology of Tropical Savannas. Springer, Berlin Heidelberg New York, pp 308–335

Medina E (1986) Forests, savannas and montane tropical environments. In: Baker NR, Long SP (eds) Photosynthesis in contrasting environments. Elsevier, Amsterdam, pp 139–171

Medina E (1987) Nutrients: requirements, conservation and cycles in the herbaceous layer. In: Walker B (ed) Determinants of savannas, IUBS Monographs Series No 3, Chap. 3. IRL Press, Oxford, pp 39–65

Medina E (1993) Mineral nutrition: tropical savannas. Prog Bot 54:237–253

Medina E, Silva JF (1990) Savannas of northern South America: a steady state regulated by water-fire interactions on a background of low nutrient availability. J Biogeogr 17:403–413

Meinzer FC (1978) Observaciones sobre la distribución taxonómica y ecológica de la fotosintesis C_4 en la vegetación del nordeste de Centroamérica. Rev Biol Trop 26:359–369

Montaldo P (1977) El espectro de las tribus de gramineas de los Llanos venezolanos. Turrialba 27:175–177

Njoku E (1963) Seasonal periodicity in the growth and development of some forest trees in Nigeria. J Ecol 51:617–624

Owen TP, Thomson WW (1991) Structure and function of a specialized cell wall in the trichomes of the carnivorous bromeliad Brocchinia reducta. Can J Bot 69:1700–1706

Petit-Maire N (1984) Le Sahara, de la steppe au désert. Recherche 15:1372–1382

Piedale MTF, Long SP, Junk WJ (1994) Leaf and canopy photosynthetic CO_2-uptake of a stand of Echinochloa polystachya on the Central Amazon floodplain. Are the high potential rates associated with the C_4-syndrome realized under the near-optimal conditions provided by this exceptional natural habitat? Oecologia 97:193–201

Reich PB, Borchert R (1984) Water stress and tree phenology in a tropical dry forest in the lowlands of Costa Rica. J Ecol 72:61–74

Ryan PR, Kinraide TB, Kochian LV (1994) Al^{3+}–Ca^{2+} interactions in aluminium rhizotoxicity. I. Inhibition of root growth is not caused by reduction of calcium uptake. Planta 192:98–103

Sarmiento G (1984) The ecology of neotropical savannas. Harvard University Press, Cambridge

Schmucker T, Linnemann G (1959) Carnivorie. In: Handbuch der Pflanzenphysiologie, vol XI. Springer, Berlin Göttingen Heidelberg, pp 198–283

Schuepp PH (1993) Leaf boundary layers. New Phytol 125:477–507

Solbrig OT (1993) Plant traits and adaptive strategies: their role in ecosystem function. In: Schulze E-D, Mooney HA (eds) Biodiversity and ecosystem function. Ecological studies, vol 99. Springer, Berlin Heidelberg New York, pp 97–116

Souza Moreira FM de, Silva MF da, Faria SM de (1992) Occurrence of nodulation in legume species in the Amazon region of Brazil. New Phytol 121:563–570

Stewart GR, Pate JS, Unkovich M (1993) Characteristics of inorganic nitrogen assimilation of plants in fire-prone mediterranean-type vegetation. Plant Cell Environ 16:351–363

Tieszen LL, Senyimba MM, Imbamba SK, Troughton JH (1979) The distribution of C_3 and C_4 grasses and carbon isotope discrimination along an altitudinal and moisture gradient in Kenya. Oecologia 37:337–350

von Humboldt A (1982) Südamerikanische Reise. 1808, quoted after the edition of Reinhard Jaspert, Ullstein, Berlin

von Humboldt A (1986) Ansichten der Natur JG Cotta, Stuttgart 1849, quoted after the edition of Greno Verlagsgesellschaft, Nördlingen

Vareschi V (1980) Vegetationsökologie der Tropen. Ulmer, Stuttgart

Walter H, Breckle S-W (1984) Ökologie der Erde. Bd 2. Spezielle Ökologie der tropischen und subtropischen Zonen. G Fischer, Stuttgart

Fig. 8.1A-C

Fig. 8.2.
Inselbergs along the Guayana
shield in Venezuela. (Huber
and Alarcon 1988)

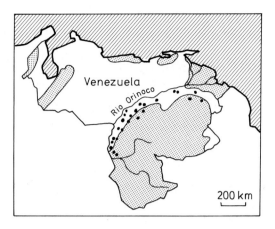

Fig. 8.3. Scheme of the vegetation of a granitic inselberg in South America. (French
Guiana). (Schnell 1987, with kind permission of Masson S. A. Paris)

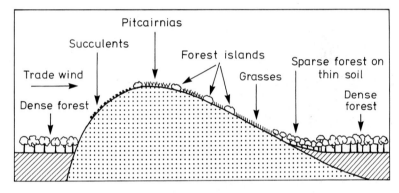

inselbergs. The former are more abundant in Africa, the latter in South
America (Barthlott et al. 1993).

Sub-islands on the rock are formed by individual plants or small groups
of plants growing in humus in cracks, gaps and hollows (Fig. **8.4**). Xero-
philic plants, such as succulents and epiphytes grow saxicolously. Patches of
shrubbery (Fig. **8.5A**) and small forests often form on top of the inselbergs
(Fig. **8.5B**). In South America these forests comprise the only deciduous
vegetation units occurring in the area around the Guayana shield, with trees

◀**Fig. 8.1A-C.** Inselbergs along the Rio Orinoco, Venezuela. A Savanna near Puerto Ayacucho at the Southern rim of the Llanos, where the Llanos border the Guayana highlands,
with inselbergs in the background. **B** Individual inselberg near Puerto Ayacucho. **C**
Eroded inselbergs, fragmented to heaps of small rocks, in the savanna, near Caicara del
Orinoco

Fig. 8.4A, B.
Isolated plants on inselbergs:
Pitcairnia pruinosa (**A**), and
a small group of *Vellozia
tubiflora* (**B**) on inselbergs
along the Rio Orinoco near
Puerto Ayacucho, Venezuela.
(Dry season, March 1991)

Fig. 8.5A, B.
Shrubs (**A**) and deciduous
forest with the palm *Syagrus
orinocencis* (**B**) on an inselberg near Puerto Ayacucho,
Venezuela. (Dry season,
February 1989)

Fig. 8.6. Wetter sections and pools on an inselberg near Puerto Ayacucho, Venezuela, where seasonally water is flowing and small ponds with water are formed in hollows. (Dry season, February 1989)

like *Pseudobombax chrysati*, *Tabebuia orinocensis* and *Yacaranda filicifolia* (Fig. **7.15**), and also palms (*Syagrus orinocensis*, Fig. **8.5**). While most sites on inselbergs are extremely dry and hot in the dry season (see below Sects. **8.2** and **8.3**), there are also wetter sections (Fig. **8.6**) and even rock pools, which seasonally keep small ponds of water and harbour aquatic plants. Huber (1980) listed the natural habitats of the inselbergs along the Orinoco in Venezuela as belonging to the "flower paradises and botanical gardens of the earth".

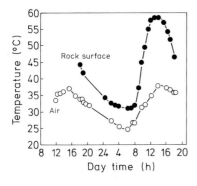

Fig. 8.7.
Temperature of the air above an inselberg (o) and the rock surface (●) during two consecutive days (15-16 March 1991) near Galipero (Pto. Ayacucho, Territoria Federal Amazonas, Venezuela)

   As this brief survey shows, plant sociology and vegetation analysis are just beginning to ask the pertinent questions to provide a deeper understanding of these exciting sites (see Barthlott et al. 1993). Ecophysiology of this fascinating vegetation is still more poorly developed, perhaps the one exception being studies of desiccation tolerance in inselberg plants. Where there is strong seasonality, pools will dry out in the dry season. The rock surface during midday may easily heat up to temperatures around 60 °C (Fig. **8.7**). Not only cryptogams but also many angiosperms of the inselbergs have been shown to be able to overcome such dry periods by ecophysiological adaptations, particularly by desiccation tolerance as described below.

8.2
Cyanobacteria

8.2.1
Ubiquity

A conspicuous feature of the inselberg rocks is their superficial appearance of dark coloration. This was first noted in Venezuela by Alexander von Humboldt (1849) who described the black surface of the rocks in the riverbed of the Orinoco and also of the rock outcrops further away:

"In the Orinoco, especially in the cataracts of Maypures and Atures, all granite blocks and even white pieces of quartz to the extent they are touched by the water of the Orinoco develop a greyish-black cover which does not penetrate more than 0.01 line[2] into the interior of the rocks. One might think to see basalt or fossils stained by granite. In fact the sheath appears to contain manganese oxide and carbon."[3]

Regarding the dark colour of the rock outcrops at a greater distance from the current river bed of the Orinoco he assumed that the river had extended much further in earlier times:

"This assumption is supported by several observations. One sees black caves 150 to 180 feet above the present water level. Their existence teaches ... that the streams, whose size presently excites our admiration, are only humble remains of the enormous amounts of water in archaic times ... These simple observations even did not escape the rough natives. Everywhere they drew our attention to the old waterlevel."[4]

---

[2] 1 line = 2.0 to 2.5 mm; Fig. **8.8**.

[3] "Im Orinoco, besonders in den Cataracten von Maypures und Atures, ... nehmen alle Granitblöcke, ja selbst weiße Quarzstücke, so weit sie das Orinoco-Wasser berührt, einen graulich-schwarzen Überzug an, der nicht 0,01 Linie ins Innere des Gesteins eindringt. Man glaubt Basalt oder mit Granit gefärbte Fossilien zu sehen. Auch scheint die Rinde in der That braunstein- und kohlenstoffhaltig zu sein."

[4] "Diese Vermutung wird durch mehrere Umstände bestätigt ... man sieht schwarze Höhlungen 150 bis 180 Fuß über dem heutigen Wasserstand erhaben. Ihre Existenz lehrt ..., daß die Ströme, deren Größe jetzt unsere Bewunderung erregt, nur schwache Überreste von der ungeheuren Wassermenge der Vorzeit sind."
"Selbst den rohen Eingeborenen ... sind diese einfachen Bemerkungen nicht entgangen. Überall machten uns die Indianer auf die Spuren des alten Wasserstandes aufmerksam."

Fig. 8.8. Black surface of the rock of an inselberg near Puerto Ayacucho, Venezuela, a small section was removed to show that only the very surface is coloured

Fig. 8.9A-C. Petroglyphs of an inselberg at Pintado near Puerto Ayacucho, Venezuela ▶ (**A**) and on granite rocks near a small pond in the secondary forest covering the US Virgin Island St. John, Lesser Antilles (**B, C**)

The finding of prehistoric petroglyphs in the rocks of the inselbergs (Fig. 8.9A) also played a role in Humboldt's discussion, since he asked how the indians might have found access to the steep walls of rock for carving them:

"Between Encaramada and Caycara on the banks of the Orinoco one frequently finds ... hieroglyphic pictures in considerable height on the rock faces, which now would only be accessible by means of extraordinarily high scaffolding. If one asks the natives how these pictures could have been carved, they reply with a smile, as if they were telling a story which only the white man may not know: that in the days of the extended waters their forefathers were boating in canoes in such a height. This is a geological dream for the solution of the problem of a long vanished civilization."[5]

---

[5] "Noch mehr: zwischen Encaramada und Caycara an den Ufern des Orinoco befinden sich häufig hieroglyphische Figuren in bedeutender Höhe auf Felsenwänden, die jetzt nur mittels außerordentlich hoher Gerüste zugänglich sein würden. Fragt man die Eingebornen, wie diese Figuren haben eingehauen werden können, dann antworten sie lächelnd, als erzählten sie eine Sache, die nur ein Weißer nicht wissen könne: 'daß in den Tagen der großen Wasser ihre Väter auf Canots in solcher Höhe gefahren seien'. Dies ist ein geologischer Traum, der zur Lösung des Problems von einer längst vergangenen Civilisation dient."

Fig. 8.9A-C

Furthermore, regarding the extensive distribution of the petroglyphs, Humboldt alludes to his correspondence with Sir Robert Schomburgk:

"I may be permitted to include a remark, which I take from a letter of the distinguished traveller Sir Robert Schomburgk: 'The hieroglyphic pictures have a much wider distribution than you might have assumed' ... the symbolic figures, which Robert Schomburgk found engraved in the river valley of the Essequibo at the rapids of Waraputa according to his observation resemble the truely caribbean ones on one of the small Virgin Islands (St. John)[6], however, notwithstanding the wide expansion of the invasions of caribbean tribes and the ancient power of this beautiful human race, I can not believe that this whole immense belt of carved rocks which cuts across a large part of South America from west to east is the work of the Caribs. They are rather traces of an ancient civilization, which possily belongs to an epoch, when the tribes which we distinguish nowadays were still unknown by name and relationship. Even the reverence, which everywhere is deferred to these rough sculptures of the forefathers, proves that the present indians have no idea of the creation of such works."[7]

However, there are also already some slight reservations in Humboldt's writings regarding the inorganic nature of the black sheets on the rocks. He stressed that it **appears** to be manganese oxide and carbon:

"I say, it appears; because the phenomenon has not been investigated diligently enough. At the Orinoco these leadlike coloured rocks if wetted emit harmful emanations. One believes their proximity to be a cause of fevers."[8]

A biological cause is suggested by the observation that the black coloration is associated with only the organically rich white-water rivers and not the more sterile black-water rivers:

"It is also noteworthy that the rivers with black water, *aguas negras*, the coffee-brown or wine-yellow waters, in South America do not stain the granite rocks black."[9]

---

[6] See Fig. **8.9 B**.

[7] "Es sei mir erlaubt, hier noch eine Bemerkung einzuschalten, welche ich einem Briefe des ausgezeichneten Reisenden Sir Robert Schomburgk an mich entnehme: 'Die hieroglyphischen Figuren haben eine viel größere Ausbreitung, als Sie vielleicht vermutet haben'. ... Die symbolischen Zeichen, welche Robert Schomburgk in dem Flußtal des Essequibo bei den Stromschnellen ... von Waraputa eingegraben fand, gleichen zwar nach seiner Bemerkung den ächt caraibischen auf einer der kleinen Jungferninseln (St. John); aber ungeachtet der weiten Ausdehnung, welche die Einfälle der Caraiben-Stämme erlangten, und der alten Macht dieses schönen Menschenschlages, kann ich doch nicht glauben, daß dieser ganze ungeheure Gürtel von eingehauenen Felsen, der einen großen Theil Südamerikas von Westen nach Osten durchschneidet, das Werk der Caraiben sein sollte. Es sind vielmehr Spuren einer alten Civilisation, die vielleicht einer Epoche angehört, wo die Racen, die wir heut zu Tage unterscheiden, nach Namen und Verwandtschaft noch unbekannt waren. Selbst die Ehrfurcht, welche man überall gegen diese rohen Sculpturen der Altvordern hegt, beweist, daß die heutigen Indianer keinen Begriff von der Ausführung solcher Werke haben."

[8] "Ich sage: sie scheint; denn das Phänomen ist noch nicht fleißig genug untersucht ... Am Orinoco geben diese bleifarbigen Steine, befeuchtet, schädliche Ausdünstungen. Man hält ihre Nähe für eine fiebererregende Ursache."

[9] "Auffallend ist es auch, daß die Flüsse mit schwarzen Wassern, *aguas negras*, die caffeebraunen oder weingelben, in Südamerika die Granitfelsen nicht schwarz färben."

Fig. 8.10A-C.
Typical X-ray fluorescence spectra of rocks in the riverbed of the Orinoco at Pto. Paez (**A**) and of an inselberg near Puerto Ayacucho (**B, C**), Venezuela. The peaks are marked by the *elemental symbols*. (By courtesy of R. Stelzer, Hannover; see Büdel et al. 1994)

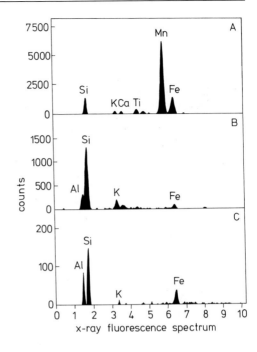

Table 8.1. Elemental analysis (elements of order-number 11-80) by X-ray fluorescence spectroscopy of samples from sandstone rocks of the Serriania Parú (04°25'N, 65°32'W, 1200 m a.s.l.), inselbergs along the Orinoco and the riverbed of the Orinoco near Puerto Ayacucho. (See Büdel et al. 1994)

Element %	Serrania Parú[a]	Inselbergs Orinoco[b]	Riverbed Orinoco[c]
Al	37.9 ± 9.5	16.8 ± 1.3	7.8 ± 0.1
Si	46.0 ± 10.0	55.1 ± 4.2	18.9 ± 3.1
S	3.2 ± 2.4	1.1 ± 0.4	0
K	4.3 ± 0.7	7.5 ± 1.2	1.9 ± 0.3
Ca	1.1 ± 0.7	2.7 ± 1.0	2.8 ± 0.2
Mn	0.2 ± 0.1	0.5 ± 0.1	49.7 ± 4.0
Fe	0.5 ± 0.3	6.3 ± 1.3	15.5 ± 0.9
Others	6.8 ± 0.5	10.0 ± 2.4	3.4 ± 0.3

[a] Serrania Parú 3 samples, 1-2 analyses each.
[b] Inselbergs Orinoco 8 samples, 4–13 analyses each.
[c] Riverbed Orinoco 1 sample, 10 analyses.
  Values are $\bar{x}$ ± SE for the averages of the individual analyses of the three and eight samples in [a] and [b] respectively, and for the ten analyses of the sample in [c].
  One sample of [b] had 82 % Os.

Fig. 8.11. The black rocks in the river bed of the Rio Orinoco near the confluence with the Rio Meta, Venezuela

Fig. 8.12. Lichens *(Peltula tortuosa)* on the rock surface of a granite inselberg near Puerto Ayacucho, Venezuela

Indeed, an elemental analysis of the black cover of the inselbergs along the Orinoco using x-ray fluorescence spectroscopy shows that there are traces of manganese only (Fig. **8.10**, Table **8.1**). There is only one exception, and these are the rocks directly in the riverbed of the Orinoco (Fig. **8.10 A**, Fig. **8.11**, Table **8.1**), where the analysis indeed indicates the dominant presence of manganese oxide. The high levels of Al and Si in all cases, of course, are due to the bed rock. The dominance of elements like S, K and Ca is consistent with the occurrence of life in these crusts. Indeed, lichens and small

Fig. 8.13A, B.
Closeup photographs of rocks covered with cyanobacteria on an inselberg near Puerto Ayacucho, Venezuela (**A**) and on the granite in the Sierra Maigualida, Guayana Highlands (05°30'N, 65°15'W, 2040 m a.s.l.) Venezuela (**B**)

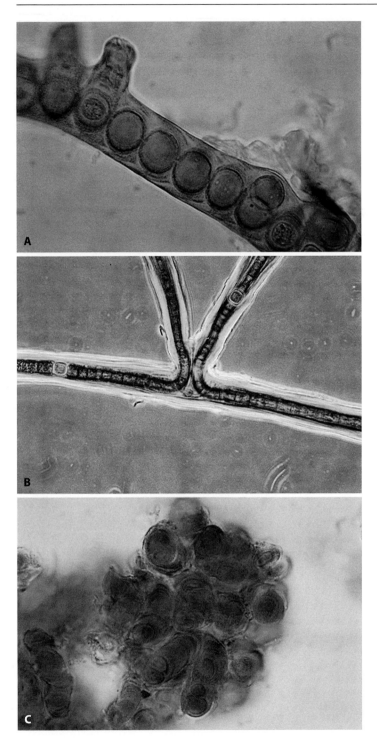

Fig. 8.14A-C

mosses are often readily discerned on the rock surfaces (Fig. **8.12**). How-ever, microscopic inspection shows that even the smooth black covers of these rocks result from living organisms. They are mainly composed of epi-lithic and endolithic cyanobacteria, predominantly of the genera *Gloeo-capsa*, *Stigonema* and *Scytonema* (Figs. **8.13** and **8.14**). Similar coverings of cyanobacteria are found on rocks throughout the tropics. The phenomenon has been described for the sandstone rocks near Cumana in eastern Vene-zuela (Golubic 1967), and examples are given in Fig. **8.15** of the granite rocks of Sierra Maigualida and sandstone rocks of Tepuis (Sierrania Parú) in the Guayana highlands of Venezuela. They resemble the "Tintenstrich" ("ink strip") formation frequently found in the European Alps, particularly on calcareous rocks (Jaag 1945).

In the tropics almost every free surface on rocks is covered by **cyanobac-terial mats and crusts**. Based on the large extension of supporting rocks these cyanobacteria overall must constitute an **enormous biomass in the tropics**. By the possession of **heterocytes** they are characterized as **N$_2$-fixing organisms**, because heterocytes are special cells in the coenobial colonies or filaments bearing the enzymatic machinery for the reduction of atmo-spheric N$_2$ (see Sect. **7.3.1.3** and box **7.5**). Thus, cyanobacterial crusts and mats possibly contribute considerably to the N-input into the ecosystems of the inselbergs themselves and via run-off into the surrounding savannas or forests. On the rock habitat they probably provide an essential starting point for succession.

8.2.2
Success on Bare Substratum

Like lichens, cyanobacterial mats and crusts are desiccation tolerant. **Wet-ting and drying cycles** have been studied in much detail for **lichens** in arid habitats. Photosynthesis of lichens is related in a complex and delicate way to the transient water conditions of the thallus (Fig. **8.16**). At very low water content, i.e. below 20 % related to dry weight, the lichens are metabolically dormant showing neither photosynthesis nor respiration. Between 20 and 50 % water content, photosynthetic net CO$_2$-uptake increases sharply and then reaches a plateau as optimal water content is attained. However, when the water content increases further and thalli are fully saturated with water, CO$_2$-assimilation is depressed. This is due to increased limitation of photo-synthesis by CO$_2$ diffusion when the capillary system of the lichen thallus is infiltrated. Thus, upon drying the assimilation rates may first increase again

◄**Fig. 8.14A-C.** Cyanobacteria composing the black crusts of the inselbergs along the Ori-noco, Venezuela. **A** *Stigonema ocellatum* (Dillw.) Thur. ex Born. et Flah. **B** *Scytonema crassum* Naeg. in Kuetz. **C** *Gloeocapsa sanguinea* (Ag.) Kuetz. em. Jaag. (Courtesy B. Büdel, Kaiserslautern; see Büdel et al. 1994)

Fig. 8.15A-C. Black granite rocks of the Sierra Maigualida (at 05°30'N, 65°15'W, 2040 m a.s.l.; **A** and **B**, in **B** shortly after a rain storm) and sandstone rock of the Serrania Parú (at 04°25'N, 64°32'W, 1200 m a.s.l.; **C** with cyanobacteria). Guayana Highlands Venezuela, March 1991

Fig. 8.16.
Net CO_2 uptake in the light (o) and net CO_2-release in the dark (●) of the lichen *Ramalina maciformis* at varied thallus water content related to dry weight. (Lange 1988, with kind permission of the author and J of Ecology)

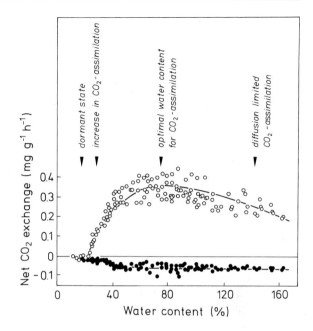

and then decline as the thalli desiccate (Lange 1988). It can readily be deduced that this similarly applies to inselberg lichens and also to cyanobacterial crusts and mats.

The role of cyanobacteria in **cryptogamic soil crusts of deserts** and other dry habitats has recently received some attention (Lange et al. 1992; Evans and Ehlehringer 1993; Jeffries et al. 1993a,b). Almost nothing is known, however, about the ecophysiology of cyanobacteria on rocks in the tropics, and in view of the enormous distribution of terrestrial cyanobacteria, which cover almost any surface not occupied by other vegetation, this is quite astonishing. The major reasons for the **success of cyanobacteria on the bare substratum of rocks** appears to be

- **desiccation tolerance,**
- a **potential to adapt to high light intensities,** and
- the **ability to fix atmospheric dinitrogen.**

Recovery from **desiccation** may occur within a few minutes up to several hours after rewetting (Jones 1977; Coxson and Kershaw 1983; Lüttge et al. 1995). It depends on the duration of the dormant state of desiccation, and frequent drying and wetting cycles maintain stability (Scherer and Zhong 1991). The sequence of events during recovery is firstly, reappearance of respiration, followed by photosynthesis and finally N_2-fixation (Scherer et al. 1984). Fig. **8.17** shows the loss and reappearance of chlorophyll-

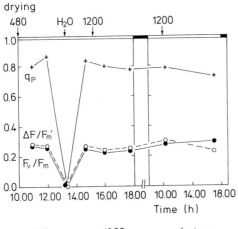

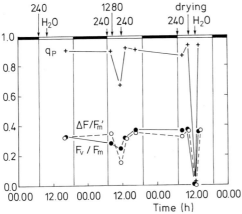

Fig. 8.17.
Optimal quantum yield of photosystem II (F_v/F_m), effective quantum yield ($\Delta F/F_m'$) and photochemical quenching (q_p) of cyanobacterial crusts of an inselberg near Seguéla, Ivory Coast (07°42'N, 06°43'W) in drying (*arrow drying above the graphs*) and rewetting (*arrow H_2O above the graphs*) cycles and during transfers between lower and higher light intensities (*arrows with numbers above the graphs* giving light intensities at $\lambda = 400\text{-}700$ nm in µmol photons m^{-2} s^{-1}). Dark and *white bars above the graphs* indicate dark and light periods respectively. (Lüttge et al. 1995)

fluorescence signals in drying and rewetting cycles performed in the laboratory on a rock sample with cyanobacterial crusts from an inselberg in Ivory Coast with the cyanobacteria *Stigonema mamillosum* (Lyngb.) Ag., *Scytonema lyngbioides* Gardner and *Gloeocapsa sanguinea* (Ag.) Kütz. emend Jaag, and traces of *Stigonema ocellatum* Thuret. It indicates that photosynthetic activity is fully restored within 60 min after rewetting following desiccation. Similar recovery times of fluorescence yield were obtained with lichens having cyanobacteria as the phycobionts (Lange et al. 1989).

The response of cyanobacteria to **high light intensities** depends greatly on the irradiance experienced during growth. Cells grown at 50 µmol photons m^{-2} s^{-1} (at $\lambda = 400\text{--}700$ nm) or below are already photoinhibited at 250 µmol m^{-2} s^{-1} and strongly affected at still higher light intensities (Samuelsson et al. 1985; Lüttge et al. 1995). However, inselberg rocks usually receive full sunlight unless clouds and rain quench exposure. Fig. **8.17** shows that in

Fig. 8.18 A-C.
Potential quantum yield of photosystem II after dark adaptation (F_v/F_m), effective quantum yield ($\Delta F/F_m'$), photochemical quenching (q_p) and relative photosynthetic electron transport rates ($\Delta F/F_m' \times PPFD$) related to photosynthetic photon fluence density (*PPFD*) as determined in the laboratory with tropical rock samples. **A** and closed symbols in **C** sample of a granitic rock from the eastern slope of the coastal mountain range in the SE of Madagascar (24°49'S, 46°57'E, 100 m a.s.l.) with *Stigonema minutum* (Ag.) Hassal and *Gloeocapsa magma* (Bréb.) Hollerbach. **B** and open symbols in **C** sample of rock outcrops in Menagesha State Forest near Addis Ababa, Ehtiopia (09°04'N, 38°22'E, 2800 m a.s.l.) with *Gloeocapsa sanguinea* (Ag.) Kütz. (Lüttge et al. 1995)

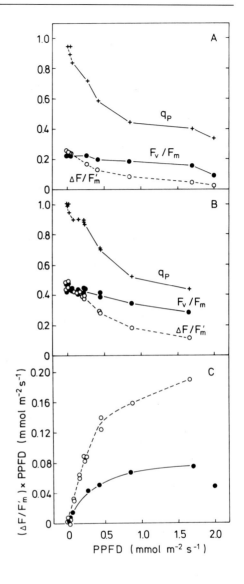

the inselberg sample from Ivory Coast, a transfer from 480 to 1200 μmol photons m^{-2} s^{-1} did not affect fluorescence yield and photochemical fluorescence quenching, indicating unimpaired photochemical and carbon-assimilatory activity, while a transfer from 240 to 1280 μmol photons m^{-2} s^{-1} resulted in a slight and rapidly reversible inhibition. As in higher plants (Schreiber and Bilger 1993), potential quantum yield of photosystem II after dark adaptation (F_v/F_m), effective quantum yield ($\Delta F/F_m'$) and photochemical fluorescence quenching (q_p) (see Sect. **3.6.2.4**, Box **3.8**) decrease with increasing light intensity in cyanobacterial crusts (Fig. **8.18**). The multiplication of $(\Delta F/F'_m)$ by photosynthetic photon fluence density (PPFD) gives relative photosynthetic electron transport rates, which saturated at the high intensities of 1000 μmol photons m^{-2} s^{-1} or above in the inselberg sample from Madagascar and the rock outcrop sample from Ethiopia measured in the experiments of Fig. **8.18**. Thus, although even cyanobacteria crusts grown under full sun exposure may be subject to partial photoinhibition, it is quite clear, that cyanobacteria can adapt very well to very high irradiance (Lüttge et al. 1995).

The third important trait of cyanobacterial rock crusts highlighted above, i.e. **N₂-fixation**, needs little additional comment, since it represents such an essential nutritional prerequisite.

8.3
Vascular Plants

8.3.1
Diversity and Life Forms

As mentioned in the introduction, the floristic diversity of cryptogamic and phanerogamic vascular plants of inselbergs is quite large (Sect. **8.1**). An analysis of granitic rock vegetation in Brazil shows that also the diversity of **life forms** occurring on rock outcrops is quite noticeable (Table **8.2**). It is interesting that among the life forms distinguished by Raunkiaer (see Sect. **3.4**) the **hemicryptophytes** are by far the most outstanding group, i.e. plants having rosettes, tillers and rhizomes on the ground with only parts of the plants protruding. Little work is available on the ecophysiological significance of this particular life form for the stress experienced on tropical rock

Table 8.2.
Relative abundance of various life forms of vascular plants on granitic rocks in Brazil (State of Rio de Janeiro and adjacent regions). (Courtesy S.T. Meirelles, A.C. da Silva and E. A. de Mattos.) For the description of Raunkiaer's life forms see Sects. **3.4** and **3.4.1**. For the other life forms cf. also Chap. **4**

Life form		Relative abundance (%)
Mega-		0
Meso-	phanerophytes	0
Micro-		8
Nano-		11
Chamaephytes		1
Hemicryptophytes		32
Cryptophytes		5
Terophytes		5
Lianas		1
Succulents		4
Epiphytes		3
Hemiepiphytes		1
Tank-formers		5
Epilithic plants		6
Atmospheric plants		6
Hygroscopic plants		12

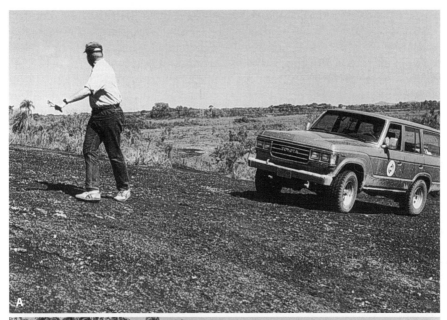

Fig. 8.19A, B. Negotiation of an inselberg near Puerto Ayacucho, Venezuela, with heavy equipment (**A**) reaching the forest of deciduous trees and the palm *Syagrus orinocensis* near the top (**B**). Dry season, March 1991

surfaces. Most likely, similar to deserts where such life forms are also frequent, any adaptive value in inselbergs may also be related to the water factor. Although at least some inselbergs are readily accessible even with heavy equipment (Fig. **8.19**) only very little ecophysiological work has been performed on site. However, from the frequency and intense development of species of Agavaceae, Bromeliaceae, Cactaceae and also Orchidaceae we may easily conclude that **crassulacean acid metabolism** (CAM) is an important mode of photosynthesis on these rock habitats. On the other hand, **C_3-photosynthesis** is also dominant. On the inselbergs along the Orinoco both CAM-bromeliads (e.g. *Ananas ananassoides, Bromelia goeldiana*, and the epiphytic *Tillandsia flexuosa*) and C_3-bromeliads (*Pitcairnia armata, P. bulbosa, P. pruinosa*) occur together. **C_3/CAM-intermediate Clusiaceae** (*Clusia* spec. and *Oedematopus* spec., see Sect. **4.4.1**) are found on South-American inselbergs. Bromeliads, like *Ananas ananassoides* and *Bromelia goeldiana*, develop the contrasting phenotypes of yellow-reddish fully exposed plants and dark-green shaded plants under the canopy of shrubbery and small forests respectively, as already described for *Bromelia humilis* (Sect. **3.6.2.2**).

8.3.2
Desiccation Tolerance

Desiccation tolerance appears to be the most outstanding particular ecophysiological adaptation among the plants of tropical inselbergs. Drought tolerance allows plants to overcome shorter periods of stress, such as the duration of dry seasons. In contrast, desiccation tolerant plants can survive equilibration with ambient air humidity below 50 % and down to 0 %, and withstand the loss of more than 90 % of their normal water content for many years (Gaff 1977, 1987). Desiccation tolerant plants are **poikilohydric** in contrast to the homoiohydric non-desiccation tolerant plants. They are also called **resurrection plants** because of their recovery from dryness.

Desiccation tolerant vascular plants, especially phanerogamic species, were initially known especially from Africa, where they comprise the most important component of inselberg communities (Gaff 1977). More recently they also have been seen to play a considerable role in South America, as well as being found in Australia and India (Gaff 1987; Meirelles et al. 1994). Among the phanerogams desiccation tolerant species mostly are monocotyledons and there are fewer desiccation tolerant dicotyledons (Gaff 1977), with the latter apparently absent from South America (Gaff 1987). Although among cryptogamic vascular plants desiccation tolerance occurs in a high proportion of taxa (e.g. in the class Lycopodiopsida, order Selaginellales; in the class Pteridopsida order Schizales and Pteridales), tolerance appears to be less extreme than observed in angiosperms. Ferns and fern allies mostly grow in positions where they are largely protected from direct sunlight, while angiosperm desiccation-tolerant plants are often fully exposed (Gaff 1977). In South America the most important desiccation tolerant families

Fig. 8.20. *Vellozia squamata* in the cerrados near Brasilia, Brazil

are the monocotyledons Velloziaceae (Figs. **8.4B, 8.20**), Poaceae and Cyperaceae (Meirelles et al. 1997). In fact, all Velloziaceae so far studied are desiccation tolerant and there are American and African genera:

- American, *Vellozia*;
- African, *Xerophyta*, *Talbotia*.

There are, of course, conspicuous **changes during desiccation.** Plants loose their turgor and leaves shrink and may change their colour due to loss of pigments (Bewley 1979). In relation to photosynthetic pigments one can distinguish between

- **homoiochlorophyllous** plants which largely keep their pigments during desiccation, only undergo few structural modifications of their chloroplasts, and retain their photosynthetic apparatus in a recoverable form, and
- **poikilochlorophyllous** plants which loose all of their chlorophyll and 70 to 80 % of their carotenoids (xanthophylls and β-carotene) and the internal structure of their chloroplasts (thylakoids) only retaining the outer envelope.

Desiccation and recovery has been studied more recently in some detail in the poikilochlorophyllous Velloziaceae *Xerophyta scabrida* (Pax) Th. Dur. et Schinz growing in the Uluguru Mountains in Tanzania at 650 m a.s.l. (about 05°30'S, 35°30'E), where they form a semi-desert like bush vegetation on cliffs.

During desiccation respiration continued beyond the 24th hour at water potentials much lower than -3.2 MPa. An active respiration, which lasts until the end of the desiccation period, is responsible for the metabolic degradation of chlorophylls and other components of thylakoids (Tuba et al. 1996). Water uptake and rehydration after rewetting initially must occur predominantly via the surface of the leaves, because functional roots are lost during desiccation. As in other desiccation tolerant plants (Gaff 1977), water uptake subsequently occurs via the roots when new adventitious roots are developed (Tuba et al. 1993a).

The **events occurring** in *X. scabrida* **upon rehydration** a summarized in Table **8.3** after the work of Tuba et al. (1993a,b, 1994). The sequence in time

Table 8.3. Recovery of the poikilochlorophyllous desiccation tolerant Velloziaceae *Xerophyta scabrida* upon rehydration. The graphs in the lower three panels indicate the relative inreases of the structures and functions described in the left column. Compiled after data in Tuba et al. (1993a, b, 1994)

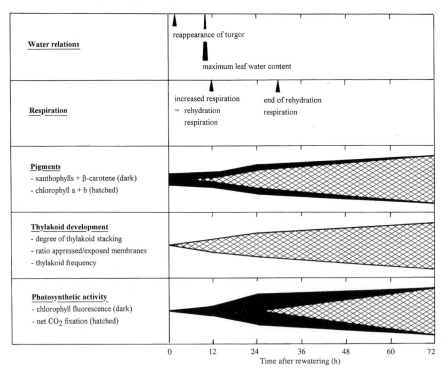

is **turgor** $\rightarrow$ **maximum leaf water content** $\rightarrow$ **respiration** $\rightarrow$ **resynthesis of carotenoids and** $\rightarrow$ **chlorophylls accompanied by thylakoid development** $\rightarrow$ **chlorophyll fluorescence yield** $\rightarrow$ **net CO_2-fixation.** Turgor reappears after 2 h, maximum leaf water content is reached after 10 h. Respiration recovers early . Mitochondrial membranes are better preserved during desiccation than the thylakoids of chloroplasts. Under stress due to chilling and freezing, heat, drought, salinity etc. membranes are often protected by compatible solutes (see Sect. **5.3**, Box **5.1**) and by changes of lipid composition. Little is known however about such protective mechanisms in desiccation tolerant plants. Respiration is increased above the normal rates after 12 h. This so-called rehydration respiration must be related to repair mechanisms and ceases after 30 h, when normal rates are attained again. In contrast to chromoplasts and gerontoplasts of senescing leaves, which cannot regreen, the chloroplasts of desiccation tolerant plants, which lose their entire photosynthetic apparatus, i.e. thylakoid membranes and pigments during desiccation, are totally rebuilt after rehydration. Tuba et al. (1993b) have named these plastids **desiccoplasts.** Some of the carotenoids, i.e. 22-28 %, are preserved during desiccation, and they may play an essential role in reorganization. Reaccumulation of carotenoids and chlorophyll a + b starts after 12 h when thylakoid development also begins to appear, as indicated by the thylakoid frequency, thylakoid stacking and the ratio appressed/exposed membranes. We recall that appressed thylakoid regions and the membranes stacked in the grana are the sites of photosystem II (see Box **3.4**). Chlorophyll fluorescence reappears at about the same time, while it takes considerably longer, i.e. about 24 h for the onset of net CO_2-uptake. After 72 h all functions have reached their normal levels again.

8.3.3
Heat Shock Proteins

There is considerable work currently being performed on the protein-biochemistry and molecular biology of heat shock proteins, a special class of polypeptides appearing after short-term heating, but also other kinds of stress. Studies with tropical cereal grasses (*Sorghum bicolor* and *Pennisetum glaucum*) have also shown, that the rapidity and extent that temperature is raised each day can lead to synthesis of special heat shock proteins, for which a detailed function is not yet known (Howarth 1991). There is no information at all about similar responses of inselberg plants, although they are likely to show this phenomenon due to the rapid and drastic changes in temperature experienced daily. On an inselberg at the Orinoco two leaves of *Ananas ananassoides* and a leaf of *Pitcairnia pruinosa* were observed to heat up by 20 to 24 °C within 5 to 9 h (Fig. **8.21**). In cyanobacteria "water-stress proteins" have been detected (Scherer and Potts 1989). The molecular biology of desiccation tolerance is currently advanced, in particular with work using *Craterostigma plantagineum* (Bartels and Nelson 1994).

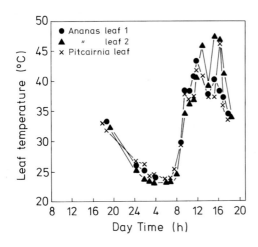

Fig. 8.21.
Temperature of two leaves of *Ananas ananassoides* (●, ▲) and a leaf of *Pitcairnia pruinosa* (×) on 15/16 March 1991 on an inselberg near Galipero (Pto. Ayacucho, Territoria Federal Amazonas, Venezuela). For air and rock temperature on the same day see Fig. **8.7**

References

Bartels D, Nelson D (1994) Approaches to improve stress tolerance using molecular genetics. Plant Cell Environ 17:659–667

Barthlott W, Gröger A, Porembski S (1993) Some remarks on the vegetation of tropical inselbergs: diversity and ecological differentiation. Biogeographia 69:105–124

Bewley JD (1979) Physiological aspects of desiccation tolerance. Annu Rev Plant Physiol 30:195–238

Bornhardt W (1900) Zur Oberflächengestaltung und Geologie Deutsch-Ostafrikas. Reimer, Berlin

Büdel B, Lüttge U, Stelzer R, Huber O, Medina E (1994) Cyanobacteria of rocks and soils in the Orinoco region in the Guayana highlands, Venezuela. Bot Acta 107:422–431

Campbell DG, Hammond HD (1989) Floristic inventory of tropical countries. The New York Botanical Garden, New York

Coxson DS, Kershaw KA (1983) Rehydration response of nitrogenase activity and carbon fixation in terrestrial *Nostoc commune* from *Stipa-Bonteloa* grassland. Can J Bot 61:2658–2668

Evans RD, Ehleringer JR (1993) A break in the nitrogen cycle in arid lands? Evidence from $\delta^{15}N$ of soils. Oecologia 94:314–317

Gaff DF (1977) Desiccation tolerant vascular plants of Southern Africa. Oecologia 31:95–109

Gaff DF (1987) Desiccation tolerant plants in South America. Oecologia 74:133–136

Golubic S (1967) Die Algenvegetation an Sandsteinfelsen Ost-Venezuelas (Cumaná). Int Rev Hydrobiol 52:693–699

Howarth CJ (1991) Molecular responses of plants to an increased incidence of heat shock. Plant Cell Environ 14:831–841

Huber O (1980) Die Felsvegetation am oberen Orinoko in Südvenezuela. In: Reisig H (ed) Blumenparadiese und botanische Gärten der Erde. Pinguin-Verlag, Innsbruck und Umschau Verlag, Frankfurt, pp 200–203

Huber O, Alarcon C (1988) Mapa de vegatacion de Venezuela, Ministerio del Ambiente y de los Recursos naturales renovables, Caracas

Jaag O (1945) Untersuchungen über die Vegetation und Biologie der Algen des nackten Gesteins in den Alpen, im Jura und im schweizerischen Mittelland. Beitr Kryptogamenflora Schweiz 9:1–560

Jeffries DL, Link SO, Klopatek JM (1993a) CO_2 fluxes of cryptogamic crusts. I. Response to resaturation. New Phytol 125:163–173

Jeffries DL, Link SO, Klopatek JM (1993b) CO_2 fluxes of cryptogamic crusts. II. Response of dehydration. New Phytol 125:391–396

Jones K (1977) The effects of moisture on acetylene reduction by mats of blue-green algae in sub-tropical grassland. Ann Bot 41:801–806

Lange O (1988) Ecophysiology of photosynthesis:performance of poikilohydric lichens and homoiohydric mediterranean sclerophylls. J Ecol 76:915–937

Lange OL, Bilger W, Rimke S, Schreiber U (1989) Chlorophyll fluorescence of lichens containing green and blue-green algae during hydration by water vapor uptake and by addition of liquid water. Bot Acta 102:306–313

Lange OL, Kidron GJ, Büdel B, Meyer A, Kilian E, Abeliovich A (1992) Taxonomic compostion and photosynthetic characteristics of the 'biological soil crusts' covering sand dunes in the western Negev Desert. Funct Ecol 6:519–527

Lüttge U, Büdel B, Ball E, Strube F, Weber P (1995) Photosynthesis of terrestrial cyanobacteria under light and desiccation stress as expressed by chlorophyll fluorescence and gas exchange. J Exp Bot 46:309–319

Meirelles ST, de Mattos EA, da Silva AC (1997) Potential desiccation tolerant vascular plants from southeastern Brazil. Pol J Env Sci (in press)

Samuelsson G, Lönneborg A, Rosenqvist E, Gustafsson P, Öquist G (1985) Photoinhibition and reactivation of photosynthesis in the cyanobacterium *Anacystis nidulans*. Plant Physiol 79:992–995

Scherer S, Potts M (1989) Novel water stress protein from a desiccation-tolerant cyanobacterium. Purification and partial characterization. J Biol Chem 264:12546–12553

Scherer S, Zhong ZP (1991) Desiccation independence of terrestrial *Nostoc commune* ecotypes (Cyanobacteria). Microb Ecol 22:271–283

Scherer S, Ernst A, Chen T-W, Böger P (1984) Rewetting of drought-resistant blue-green algae:time course of water uptake and reappearance of respiration, photosynthesis, and nitrogen fixation. Oecologia 62:418–423

Schnell R (1987) La flore et la végétation de l'Amérique tropicale. Masson, Paris

Schreiber U, Bilger W (1993) Progress in chlorophyll fluorescence research:major developments during the past years in retrospect. Prog Bot 54:151–173

Steyermark JA (1977) Future outlook for threatened and endangered species in Venezuela. In: Prance GT, Elias TS (eds) Extinction is forever. The New York Botanical Garden, New York, pp 128–135

Tuba Z, Lichtenthaler HK, Csintalan Z, Pócs T (1993a) Regreening of desiccated leaves of the poikilochlorohyllous *Xerophyta scabrida* upon rehydration. J Plant Physiol 142:103–108

Tuba Z, Lichtenthaler HK, Maroti I, Csintalan Z (1993b) Resynthesis of thylakoids and functional chloroplasts in the desiccated leaves of the poikilochlorohyllous plant *Xerophyta scabrida* upon rehydration. J Plant Physiol 142:742–748

Tuba Z, Lichtenthaler HK, Csintalan Z, Nagy Z, Szente U (1994) Reconstitution of chlorophylls and photosynthetic CO_2 assimilation upon rehydration of the desiccated poikilochlorohyllous plant *Xerophyta scabrida* (Pax) Th. Dur. et Schinz. Planta 192:414–420

Tuba Z, Lichtenthaler HK, Csintalan Z, Szente K (1996) Loss of chlorophylls, cessation of photosynthetic CO_2 assimilation and respiration in the poikilochlorohyllous plant *Xerophyta scabrida* during desiccation. Physiol Plant 96:383–388

von Humboldt A (1849) Ansichten der Natur. JG Cotta, Stuttgart. Quoted after the edition of Greno Verlagsgesellschaft, Nördlingen 1986

CHAPTER 9

Páramos

9.1
Summer Every Day, Winter Every Night

The cold tropics (Sect. **1.2**) comprise the "regions within the tropics occur-ring between the upper limit of continuous, closed-canopy forest (often around 3500–3900 m) and the upper limit of plant life (often around 4600–4900 m)". In this way Rundel et al. (1994a) define "tropical alpine environments" in their recent book. They use "alpine" as a more general term in an attempt to avoid regional terms like *páramo* and *jalca* for the moist Andes and *puna* for the drier Andes in South America and afro-alpine and moorland in Africa. However, "alpine" is also a regional term applying to environments outside the tropics. On the other hand, since the conditions and the physiognomy of vegetation are similar on tropical mountains in different continents, especially in Africa and South America, we might as well choose the term páramo. Increasingly, this is used as the general term to describe vegetation in the cold tropics extending from somewhat above 3000 m to nearly 5000 m above sea level (Fig. **9.1**).

The high altitude tropical environments were succinctly described by Hedberg's (1964a) aphorism "summer every day and winter every night". The most important feature of the tropical alpine zone is the "**Frostwechselklima**"[1] (Troll 1943) with an extraordinarily high amplitude of day/night fluctuations of humidity and especially of temperature (Fig. **9.2**). Clearly, general characteristics of tropics are strongly accentuated in these high altitudes, such that **daily oscillations of temperature** are much more pronounced than the seasonal ones. Thus, nocturnal frosts followed by high day-time temperatures represent the most conspicuous stress to which plants are exposed in these environments. Additional, often very important stressors, are limited water supply and mineral nutrition.

---

[1] day-night freezing climate

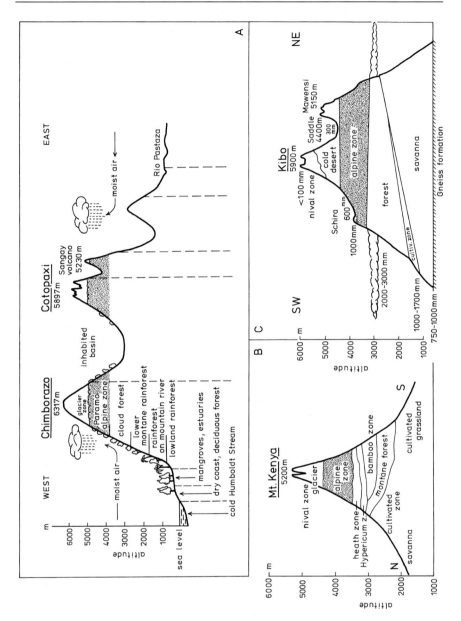

Fig. 9.1A-C. Profiles of high tropical mountains in South America (**A** Chimborazo and Cotopaxi of the Andes) and Africa (**B** Mount Kenya; **C** Mt. Kilimanjaro) with the altitudinal vegetation zones indicated, and in **C** also the approximate annual precipitation. (Walter and Breckle 1984, with kind permission of S.-W. Breckle and G. Fischer-Verlag)

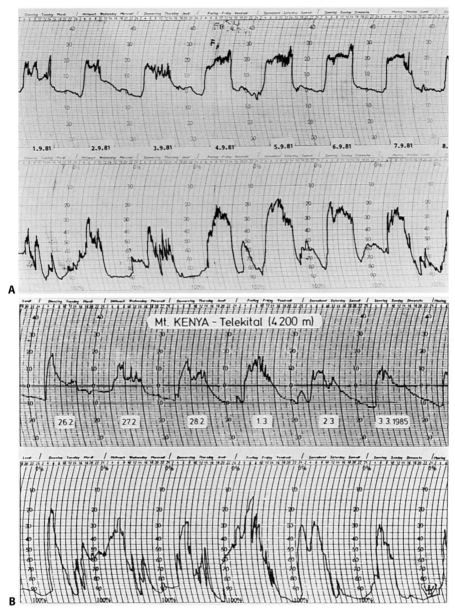

Fig. 9.2 A, B. Thermohygrograms obtained in September 1981 at the Shira Plateau on Mt. Kilimanjaro at about 3950 m (**A**) and in March 1985 on Mt. Kenya at 4200 m altitude (**B**). *Upper part* of each graph temperature °C; *lower part* relative humidity %. (Courtesy E. Beck)

9.2
The Stress Factor Frost

With temperatures below 0 °C every night, frost is the permanently dominating environmental stress factor ("stressor") in the tropical alpine habitat.
Some relations between **cold stress** and **cold resistance** are presented in Fig.
9.3 according to the stress concept (see Sect. **3.3.1.2**; Box **3.1**).

First of all we need to distinguish between **low-temperature** stress at temperatures above the freezing point (0 to +6 °C) and stress caused by **sub-
freezing temperatures**. Low-temperature stress may lead to a loss of fluidity
of membrane lipids or to an increase in membrane rigidity with many consequences for membrane permeability and intracellular compartmentation.
It also slows down many metabolic reactions. It may cause injury, elastic
and plastic strain, and it requires **chilling resistance**. It is largely presented

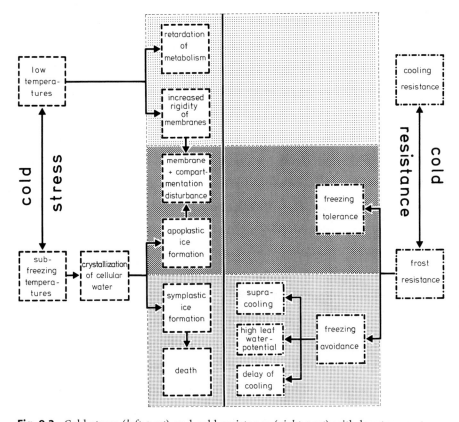

Fig. 9.3. Cold stress (*left part*) and cold resistance (*right part*) with low temperatures
above the freezing point (*upper panel*) and subfreezing temperatures (*lower panels*) as
stress factors. The terminology presented in this scheme according to the stress concept
(see Box **3.1**) is used in the present Chap. to discuss adaptations of páramo plants

by the upper panel in Fig. **9.3** but will not be discussed here further, since we are dealing with sub-freezing temperatures in the páramo habitat. Sub-freezing temperatures sooner or later may lead to **ice formation**, i.e. the crystallization of cellular water. For stress resistance, the location of ice crystal formation in the cells is critical. If it is on the outer face of the cell walls, i.e. **apoplastic**, ice formation is tolerable and thus **frost resistance** may be achieved through **freezing tolerance**. However, if ice crystals are formed in the cell interior, i.e. **intracellularly**, this always leads to cell death, and frost resistance can only be achieved by **freezing avoidance**. These two cases are represented by the lower two panels in Fig. **9.3**. They constitute options with different advantages and disadvantages, and it is interesting to note that **afro-alpine species** commonly tolerate extracellular **freezing** while **andean species** apparently rely on the **freezing avoidance** mechanism, as will be shown below in sections **9.4.1** and **9.4.2** respectively.

9.3
Life Forms of Páramo Plants

The five major life forms of páramos are

- giant-rosette-plants,
- tussocks of grasses or sedges,

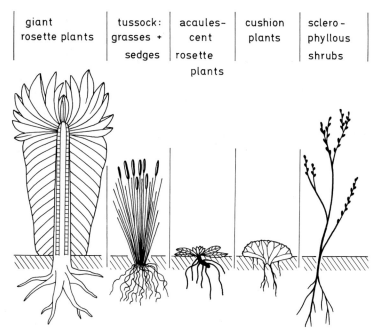

| giant rosette plants | tussock: grasses + sedges | acaules-cent rosette plants | cushion plants | sclero-phyllous shrubs |

Fig. 9.4. The five major life forms of Afro-alpine vegetation (Hedberg 1964b)

- acaulescent rosette plants,
- cushion plants, and
- sclerophyllous shrubs

(Hedberg 1964a, Beck et al. 1983) (Fig. **9.4**), and, in South America,

- cacti

may also be included.

9.3.1
Giant-Rosette Plants

The most typical life form of alpine tropical regions are the giant-rosette plants of the genera *Lobelia* (Lobeliaceae) and *Senecio* (Asteraceae) in Africa, *Espeletia* (Asteraceae) in South America and *Argyroxiphium* (Asteraceae) in Hawai'i (Fig. **9.5**). They may reach heights of several meters and have developed a number of morphological and anatomical features adapting them to the diurnal type of climate in their habitat.

The tallest plant of *Senecio johnstonii* ssp. *cottonii* actually measured by Beck et al. (1983) was 9.60 m, but an even taller one was also encountered at another location. The plants bear giant rosettes of living leaves at the end of their branches. Dead leaves usually cover the entire shoot in a dense layer. The dead leaves are kept for decades and perhaps even centuries, since the larger giant-rosette plants may be of considerable age. Estimations for *Senecio keniodendron* suggest that about 50 new leaves are formed annually, the stem growth rates are approximately 2.5 – 3 cm per year giving an age of 35 years per meter of unbranched stem (Beck et al. 1980). A 10 meter tall giant-rosette plant then might have an age of 350 years.

These **dead leaves** provide **heat insulation** of the stem. However, the morphology of the living leaf rosette, together with some physiological reactions, enables most important acclimatory strategies (Fig. **9.6**).

The conical leaf bud in the center of the rosette is protected by **nyctinastic movements of the adult leaves.** Thus, a "night-bud" is formed in these plants, which of course grow during the whole year and have no dormant periods, as plants in the temperate zone have in the form of winter buds. In *Espeletia schultzii* the new leaves, which had just developed from the central bud, wilted and died when the formation of the night-bud (by nocturnal closure of the rosette) was experimentally prevented (Smith 1974).

In African giant-rosette plants the leaf water may freeze apoplastically during the night (see Sect. **9.4.1** below). The closing mechanism of the inward movement of the adult leaves in these plants is based on water loss from the cell interior and the associated decrease of pressure in the whole tissue as turgor of the individual cells declines due to cellular loss of water.

Fig. 9.5A-D. Tropical alpine giant-rosette plants. *Espeletia timotensis* (A) and *E. schultzii* (B) Aguila Pass, Venezuela, 3600-4000 m a.s.l.; *Lobelia keniensis* (C) and *Senecio keniodendron* (D) Teleki valley, Mt. Kenya, at 4100 m and 4300 m a.s.l. respectively. (C and D courtesy E. Beck)

Fig. 9.6A-E. Giant rosettes of *Espeletia schultzii* (**A**), *E. moritziana* (**B**) and *E. timotensis* (**C**), Aguila Pass, Venezuela, 3600-4000 m a.s.l., and of *Lobelia deckenii* in the day position (**D**) and as the night-bud (**E**), Mt. Kilimanjaro, Tanzania, 3850 m a.s.l. (**D** and **E** courtesy E. Beck)

The opening of the night-bud is due to instantaneous resorption of the water and restoration of turgor after melting of the apoplastic ice in the morning.

The dense packing of the developing leaves in the central bud gives this organ a massive structure with a considerable inherent **heat-storage capacity**. In addition, excreted fluid and **mucilage** may also contribute to the heat-storage capacity. The bases of the adult leaves form tanks or cisternae where mucilageous fluid is collected during the day. In *Lobelia keniensis*, for

Fig. 9.7.
Leaf anatomy of giant-
rosette plants, with hairs
and large intercellular
spaces and mucilage
inside and on the surface
of the upper epidermins.
Cross-section of an adult
leaf of *Lobelia keniensis* in
the upper cisterna region.
(Beck et al. 1982)

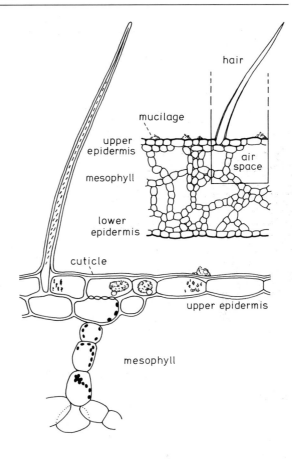

example, this may amount to several liters per rosette. During the nyctinastic leaf movements this fluid, which warms up during the day, is pressed
upwards to cover the meristematic part of the night-bud in the evening
(Beck et al. 1982).

Moreover, the leaves are very **pubescent** and have large **intercellular air
spaces** which add to the insulating effect (Fig. **9.7**). The consequences of leaf
pubescence are not straightforward. There are complex interactions
between several factors implying non-linear behaviour of the system. The
role of pubescence has been examined in more detail in *Espeletia timotensis*
in relation to temperature, wind speed and high solar radiation in the
páramo habitat (Meinzer and Goldstein 1985; Schuepp 1993). Pubescence is
more effective at high wind speeds. Increased boundary layer thickness due
to a coat of hairs hinders exchange between leaf surface and ambient wind,
and its primary effect in cool air would be an increase in surface temperature. This may be about 7 °C in the case studied, which is associated with a
small increase in transpiration ($\sim$ 17 %) due to the effects of leaf tempera-

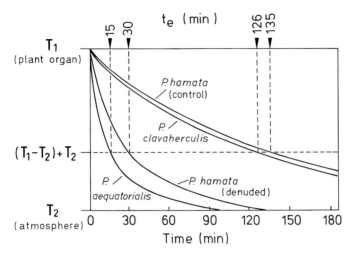

Fig. 9.8. Exponential heat decay curves and time constants, t_e, for inflorescences of *Puya hamata* (controls and denuded, respectively), *P. clava-herculis* and *P. aequatorialis*. t_e is the time when the heat decay curve crosses the line of $\frac{1}{e}(T_1-T_2) + T_2$ and is independent of the magnitude of the T_1-T_2 temperature difference. Thus, t_e gives the time it takes for a 63 % decrease in the total temperature difference between the plant organ (T_1) and the atmosphere (T_2). (After Jones 1992; Miller 1994)

ture on leaf/air water vapour pressure difference (see Sect. **3.7.2**). In numerical simulations it was shown however, that in contrast an increase of surface temperature of about 7 °C would result in a doubling of the transpiration rate of non-pubescent leaves. Increased transpiration will have a feedback on leaf temperature because of the effect of transpirational cooling, which adds to the complexity of the system.

In any event, pubescence will slow down the establishment of thermal equilibrium of plant organs with air temperature. For the pubescent inflorescences of *Puya* (Bromeliaceae) in the equatorial páramo zone of the Ecuadorian Andes this has been quantified using time constants (t_e) derived from the exponential heat decay curves (Fig. **9.8**). Clearly, in the non-pubescent species *P. aequatorialis*, which grows on rocky outcrops between 1900 and 2100 m a.s.l., t_e is much lower (15 s) than in the pubescent species *P. hamata* (135 s) and *P. clava-herculis* (126 s), and in *P. hamata* it drops to 30 s when inflorescences are denuded (Jones 1992; Miller 1994).

All of the features discussed above in this Sect. are mechanisms to delay cooling and provide freezing avoidance in the buds. Indeed, measurements in Kenya showed that only the temperature of adult leaves closely follows air temperature and is for many hours every night below the freezing point. However, nocturnal **bud temperatures** in *Lobelia* and *Senecio* are significantly higher and may even remain positive. In *Senecio brassica* for exam-

Fig. 9.9A, B.
Comparisons between leaf temperature and night-bud temperature
as related to air temperature in
Lobelia telekii (**A**) and *Senecio
brassica* (**B**) on Mt. Kenya during
the dry season in March 1979.
(Beck et al. 1982)

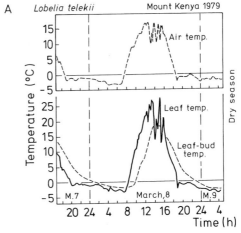

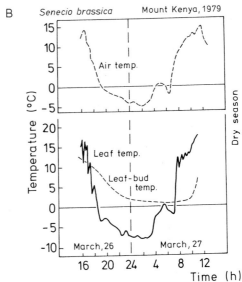

ple, at air temperatures around -8 °C, bud temperature remains +1 °C (Fig.
9.9). We shall see below (Sect. **9.4.2**) that for andean giant-rosette plants the
mechanisms of freezing avoidance based on insulation, heat storage and
delay of cooling, together with some supra-cooling effects, may suffice for
the frost-resistance of the adult leaves. However, in afro-alpine giant-rosette
plants they are insufficient and freezing tolerance is needed (Sect. **9.4.1**).

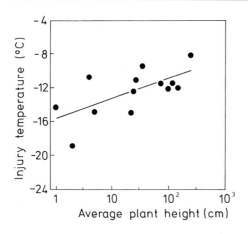

Fig. 9.10.
Injury temperature of tropical alpine plants related to average plant hight of cushions and small rosettes, shrubs and perennial herbs, giant rosettes and small trees. (Squeo et al. 1991)

9.3.2
Other Life Forms: Tussocks, Cushions, Acaulescent Rosettes, Sclerophylls

In general, it appears that smaller plants are less threatened by frost than taller ones, as suggested by Fig. **9.10**, which relates average plant height of cushions and small rosettes, shrubs and perennial herbs, giant rosettes and small trees to temperature causing injury.

In **tussock and cushion plants** the regenerating buds are insulated by adult leaves and dead material. Extensive studies on alpine plants in Europe have shown that the internal microclimate, temperature and air humidity, within such plant bodies and mats can be much different and highly protected from the outside (Reisigl and Keller 1987). The **acaulescent rosettes** (Fig. **9.11**) closely adopt the soil temperature. In the Andes, Squeo et al. (1991) found that all ground-level plants (cushions and acaulescent rosettes) were freezing tolerant. Day-night temperature changes are somewhat buffered by the heat-storage capacity of the soil and stress may be less extreme than from air temperature alone (Fig. **9.12A**). Moreover, for taller plants, a specific problem of the day-night freezing climate is not only the freezing during the night but also the process of thawing at the beginning of the day. If ice formed in leaves thaws more rapidly than ice in the stems, transpiration sets in, while water transport in the shoot is still blocked. This leads to cavitation in the xylem elements and to serious problems if it is associated with embolism.

This is avoided in acaulescent rosettes, where the whole plant body thaws at the same time. An example is given in Fig. **9.13** showing leaf temperatures and corresponding maximum photosynthetic electron transport rates at light saturation (ETR_{max}) in *Haplocarpha rueppellii* (Fig. **9.11**) during a clear day at 4100 m a.s.l. in the Simien Mountains in Ethiopia. As irradiance (PPFD in Fig. **9.13**) increased after sunrise, the rosettes thawed

Fig. 9.11A-C.
Small acaulescent rosettes of *Haplocarpha rueppellii*, Simien Mountains, Ethiopia, 4100 m a.s.l. (**A**) and *Echeveria columbiana*, Aguila Pass, Venezuela, 3600 m a.s.l. (**B**) and cushion of *Helichrysum newii*, Shira Plateau, Mt. Kilimanjaro, Tanzania (**C**). (**C** Courtesy E. Beck)

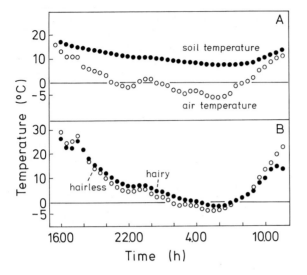

Fig. 9.12A, B.
Air temperature and soil temperature (**A**) and subepidermal temperatures (**B**) of the hairy and the hairless variety of *Tephrocactus floccosus* growing at the same site. (Keeley and Keeley 1989)

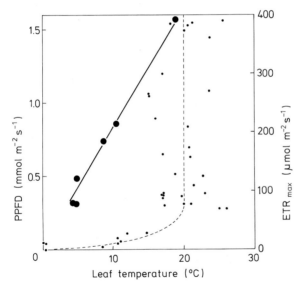

Fig. 9.13.
Relationships between leaf temperature and ambient irradiance (*PPFD; dotted line*) and maximum photosynthetic electron transport rate at light saturation (*ETR$_{max}$; solid line*) in acaulescent rosettes of *Haplocarpha rueppellii* (see Fig. **9.11A**). (Unpubl. results of the author)

and leaf temperatures increased to the average value of 20 °C when a PPFD of 400 μmol m^{-2} s^{-1} was reached. Above 400 μmol m^{-2} s^{-1} leaf temperature was independent of irradiance which clearly indicates dissipation of heat via the soil. Maximum rates of photosynthetic electron transport obtained from measurements of chlorophyll fluorescence (3. in Box **3.8 B**) at light saturation were linearly related to leaf temperature between ~ 5 and ~ 20 °C indicating limitation of photosynthesis by temperature dependence of electron transport in the thylakoids and/or biochemical reactions of CO_2-assimilation.

Some **acaulescent rosette plants** follow the **CAM** mode of photosynthesis, e.g. *Echeveria columbiana* (Crassulaceae) in the Andes (Fig. **9.11B**) (Medina and Delgado 1976). These plants thus must be able to maintain the intensive metabolism during the night as it is required for dark CO_2-fixation. They may possibly achieve this in a state of supercooling or non-ideal equilibrium freezing (see Sect. **9.4**).

In the **sclerophyllous shrubs** transpiration is reduced and water economy is sustained by leaf-xeromorphy with reduced leaf surfaces, folded leaves and other morphological/anatomical adaptations.

Rundel et al. (1994b) invoke two different reasons for the diversity of life forms found in páramo ecosystems, namely

- the complexity of environmental stresses, i. e. the diurnal freezing cycles and water and nutrient deficiencies, and
- the occurrence of microhabitat mosaics.

However, additional research is required to determine the correlations between life forms, habitat conditions, and physiological characteristics.

9.3.3
Cacti

The expression of CAM (see Box **3.11**) in andean plants has been studied in some detail in the cacti *Oroya peruviana* and *Tephrocactus floccosus* in central Peru at 4000–4700 m a.s.l. (Keeley and Keeley 1989). There was nocturnal malate accumulation even at air temperatures of -8 °C and subepidermal temperatures of -3 °C.

The occurrence of hairy and hairless varieties in *T. floccosus* offered the opportunity to demonstrate the effects of a cover by hairs for insulation (see also above Sect. **9.3.1**). During the night the hairy form had subepidermal temperatures that were several degrees higher than in the hairless form (Fig. **9.12**). A disadvantage of the hairs, however, is light scattering and therefore shading of the photosynthetically active stem tissue. Hence, the cacti have a problem of optimizing the two options to allow maximal CAM activity. It is interesting to note that at sites where one of the two forms predominated consistently the dominant form showed the higher nocturnal malate accumulation (Δmal). Thus, where the hairless type was rare, the hairy morph had the higher Δmal, and where the glabrous type was frequent, the hairy type had the lower Δmal.

9.4
Frost Resistance in Giant-Rosette Plants

9.4.1
Afro-Alpine Plants: Freezing Tolerance

As already noted above (Sect. **9.2**), formation of ice crystals in the cytoplasm is always disastrous. This is predominantly due to effects of ice competing with the ordered superficial water film of membranes. In addition ice crystals may puncture membranes and organelles. Thus, freezing tolerance is only possible with extracellular ice formation. For the afro-alpine giant-rosette plants the phenomenon has been studied in detail by E. Beck and collaborators.

For the formation of extracellular ice, the basic laws of **cell water relations** (see Box **4.1**) apply as follows:

$$\psi_{cell} = P - \pi, \tag{9.1}$$

where ψ_{cell} is the water potential of the cell, P and π are turgor and osmotic pressure respectively. During freezing the cell looses water, and ice forms outside the plasmalemma in the intercellular spaces. Thus, the protoplast shrinks and turgor becomes zero (P = 0). Therefore

$$\psi_{cell} = -\pi, \tag{9.2}$$

and at equilibrium of the protoplasts with the extracellular ice

$$\psi_{ice} = \psi_{cell} = -\pi, \tag{9.3}$$

a situation called "**equilibrium freezing**" (Fig. **9.14**). Essentially, ice forms gradually as water moves out of the symplast, and not abruptly as occurs after supercooling (see Sect. **9.4.3**). Therefore, the occurrence of nucleating agents in the apoplast, which may include mucilage (Goldstein and Nobel 1994), are important in eliciting ice-crystal formation and avoiding supercooling (Krog et al. 1979).

According to Eq. (**9.3**), the concentrations of solutes in the protoplast, which determine π, are therefore given by the water potential of the ice, which is linearly dependent on the subfreezing temperature. The increased cytoplasmic concentrations of solutes may damage or protect membranes and proteins. Often special **cryoprotective solutes** decrease the injurious effects of high ion concentrations on the membranes. In afro-alpine plants, sucrose is most likely to fulfil such a role (Beck 1994a). Cryoprotectants are similar to compatible solutes discussed above in relation to osmotic stress (Sect. **5.3** and Box **5.1**). In fact **extracellular ice formation** is nothing more than a dramatic **osmotic stress**, however, at low temperatures.

In this way, for example in *Lobelia keniensis* at -6 °C, 85 % of the tissue water is frozen. When the ice thaws in the morning, the water is immediately taken up osmotically into the cells again, and full competence of pho-

Fig. 9.14.
Equilibrium freezing: water potentials of frozen leaves (ψ_{cell} = ●) of S. *keniodendrion* and L. *keniensis* and frozen expressed cellular sap (-π = o) of S. *keniodendron* compared to ice at sub-freezing temperatures (lines: ψ_{ice} calculated by two different methods **A** and **B**, see Eq. (9.3) give the same relationships. (Beck et al. 1984)

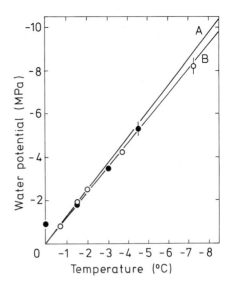

Table 9.1. Frost tolerance of leaf segments of four species of Afro-alpine megaphytes. (Beck et al. 1982)

Species	Frost tolerance (° C)	50 % damage (° C)
Senecio keniodendron	– 8	–10
Senecio brassica	–10	–15
Lobelia keniensis	Lower than –20	Lower than –20
Lobelia telekii	Lower than –20	Lower than –20

tosynthesis is regained rapidly. Using this mechanism, the afro-alpine giant-rosette plants achieve frost resistance at temperatures of -8 °C down to -20 °C (Table **9.1**).

There may also be deviations from the ideal behaviour given by Eq. (**9.3**) due to the osmotic contribution of extracellular solutes, which allow lower external water potentials (non-ideal equilibrium freezing) (Goldstein and Nobel 1991; Zhu and Beck 1991). At water losses > 50 % a matric potential is also generated which prevents intrusion of air between the cell wall and the plasmalemma, and in this way the wall may get under tension (negative turgor).

9.4.2
Andean Plants: Freezing Avoidance

In contrast to the behaviour of afro-alpine plants, andean *Espeletias* are injured by freezing of the cell water. Obviously ice formation is occurring intracellularly in these plants. They need to avoid freezing of cellular water

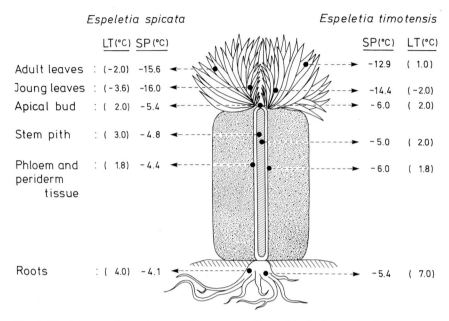

	Espeletia spicata				*Espeletia timotensis*	
	LT(°C)	SP(°C)			SP(°C)	LT(°C)
Adult leaves	: (-2.0)	-15.6			-12.9	(1.0)
Joung leaves	: (-3.6)	-16.0			-14.4	(-2.0)
Apical bud	: (2.0)	-5.4			- 6.0	(2.0)
Stem pith	: (3.0)	- 4.8			- 5.0	(2.0)
Phloem and periderm tissue	: (1.8)	- 4.4			- 6.0	(1.8)
Roots	: (4.0)	- 4.1			-5.4	(7.0)

Fig. 9.15. Scheme of a caulescent giant-rosette plant with the lowest leaf temperatures (*LT*) recorded in the field during the days of measurements and the supercooling points (*SP*) of various organs and tissues of two *Espeletia* species indicated. (Rada et al. 1985)

and survive the nights in a super-cooled state. As depicted in Fig. **9.15** for all the different organs and tissues of *Espeletia spicata* and *E. timotensis* in the Andes, the **supercooling** points, at which ice formation would happen, are always much lower than the lowest temperatures actually found in these organs and tissues. Ambient temperature is modulated in the tissues by the various strategies discussed in Sect. **9.3.1**. Supercooling and freezing avoidance for the Andean giant-rosette plants therefore must be considered as an appropriate mode of achieving sufficient frost resistance.

9.4.3
Comparison of the Strategies of Freezing Tolerance and Avoidance

If freezing tolerance and freezing avoidance are such successful adaptations of giant-rosette plants to the *Frostwechsel* climate in the African and South-American tropical high mountains, the question arises why the plants in the two continents evolved two such different modes of frost resistance. As compared to freezing and its tolerance, freezing avoidance by supercooling has one important advantage and one important disadvantage:

– the advantage is that the cells always remain metabolically competent;
– the disadvantage is the very high risk inherent in the supercooling strat-

egy, since supercooling is a thermodynamically labile state; at the lowest supercooling point the water freezes instantaneously, there is no time for water-export into the apoplast, and cell death becomes unavoidable.

Continuous metabolic competence should allow higher productivity. In the Andean páramos , where night temperatures go down to 0 °C and rarely below -5 °C, the risk of reaching the crystallization point is small (Fig. **9.15**) as compared to the benefit of higher productivity. Conversely, in the afro-alpine zone temperatures are frequently below -10 °C and the risk inherent in supercooling becomes too high to be a reasonable choice.

Hence, it appears fairly straightforward that the different nocturnal temperature regimes in the two regions led to evolution of the two different strategies, one taking the risk and obtaining more productivity and the other one avoiding the risk on account of the loss of some productivity. Indeed Table **9.2** suggests that the African *Senecio keniodendron* has a much lower annual productivity than the Andean *Espeletias* (Rada et al. 1985).

Interestingly, Lipp et al. (1994) note that in species growing at high elevations in Hawai'i features of both adaptive strategies are combined and a complete suite of characteristics with either strict tolerance or avoidance of extracellular ice formation is not expressed. Five species were studied, namely *Argyroxiphium sandwicense* and *Dubautia menziesii* (both Asteraceae), *Sophora chrysophylla* (Fabaceae), *Vaccinium reticulatum* (Ericaceae) and *Styphelia tameiameiae* (Epacridaceae) (Fig. **9.16**). Typical freezing tolerance is not fully expressed possibly due to a more recent evolutionary status of these taxa. A combination of the two possible adaptive strategies is given in that a period of supercooling occurs prior to ice nucleation. Four of the five species could tolerate extracellular ice formation to a certain degree. For example, in *S. tameiameiae* in the laboratory there was considerable supercooling prior to ice formation at -9.4°, and the latter did not cause tissue injury.

9.5
Other Stress Factors

9.5.1
Water Availability

Precipitation decreases with increasing altitude in the cold tropics (Lauer 1975; Rundel 1994). Although patterns of precipitation in páramo regions are very complex and do not give as clear a picture as the temperature patterns, it is evident that diurnal drought problems are associated with the tropical *Frostwechsel* climate. Massive structures which contribute to heat-storage capacity (Sect. **9.3.1**) may also provide water-storage capacity. Thus, the pith storage capacity of *Espeletias* in the Andes is thought to be involved in diurnal drought avoidance mechanisms. This may also explain the appar-

Table 9.2. Productivity of an African and two Andean giant-rosette plants. (After Rada et al. 1985)

	Species	Productivity (g DW m^{-2} leaf surface a^{-1})
Africa	*Senecio keniodendron*	166
South America	*Espeletia spicata*	671
	Espeletia timotensis	370

Fig. 9.16A-C. Plants of high altitudes on Haleakala volcano, Maui island, Hawai'i. *Styphelia tameiameiae* (**A**), *Dubautia menziesii* (**B**), *Argyroxiphium sandwicense* (**C**)

ent paradox that the height of giant rosette species increases with higher altitudes as this implies an increase of stem water-storage capacity (Meinzer et al. 1994). Water retaining gels have also been thought to contribute to water storage, e.g. in *Argyroxiphium sandwicense* (Carlquist 1994). Although water relations in high elevation tropical plants have been studied to some extent (Meinzer et al. 1994), ecophysiological studies of drought tolerance are less advanced than for the frost stressor.

9.5.2
Mineral Nutrition and Carbon

Nutrient relations of tropical high elevation plants have been little investigated (Rehder 1994).

An interesting case story is presented by Beck (1994b). Dead leaves kept on the stems of *Senecio*s for heat insulation (Sect. **9.3.1**) obviously are withheld from mineralization in the soil and the soil-plant nutrient cycle. However, in *Senecio keniodendron* leaves may gradually decay *in situ* on the stems, and where moisture is retained, adventitious roots emerge from the stems and produce a network in the sheath of decaying leaf bases not unlike the tank roots of bromeliads (Sect. **4.3.1**, Fig. **4.11**). Mineralised nutrients are reabsorbed by living plant tissues and the nutrient cycle is closed within the plant.

Another intriguing case of mineral nutrition is inorganic carbon acquisition from the pedosphere by terrestrial páramo species of *Isoëtes* (Lycopodiopsida, Isoëtaceae). *Isoëtes andicola* is a terrestrial species which has no stomata on the short leaves of its rosettes and gains the bulk of its carbon from the peat sediment near lakes and lagoons in the high Andes. Most of the C-uptake occurrs during the day with C-reduction via the Calvin cycle, however, there is also some CAM-type C-acquisition during the night. The significance of this pedosphere-based carbon nutrition is not clear. Sediment-based carbon acquisition is well known of aquatic *Isoëtes* species, which also may perform CAM. Among other hypotheses, in terrestrial *Isoëtes* species adaptation to seasonal droughts may be an explanation, as sealing off the leaves from the atmosphere may provide an advantage under drought-stress conditions. Other stomata-less terrestrial species with similar modes of carbon nutrition are *I. andina* and *I. novo-grandensis* in the Andes and possibly *I. hopei* in New Guinea. Aquatic *Isoëtes* species are nearly ubiquitous in páramo regions of South America providing a possible ancestral pool for the terrestrial forms with their unique strategy (Keeley et al. 1994).

9.5.3
Irradiance and Heat

Leaf hairs and scales represent an effective barrier against high-light stress, photoinhibition and damage of the photosynthetic apparatus (see also Sect. 3.7.2) and UV-radiation (Lang and Schindler 1994), as well as overheating (Melcher et al. 1994). Naturally the same insulating effect which reduces cooling during the night delays heating during the day under the *Frostwechsel* climate. Melcher et al. (1994) showed that in *Argyroxiphium sandwicense* on the Haleakala volcano on Maui island, Hawai'i, temperature of expanded leaves was similar to, or even lower than, air temperature at full solar radiation. Conversely, the apical bud in the center of the rosette was usually 25 °C warmer than air at noon. This may facilitate physiological processes required for the maintenance of growth of new leaves in the apical bud. However, this must also be the reason which limits *A. sandwicense* to altitudes of $\geq$1900 m a.s.l. below which apical bud-temperatures might reach lethal levels.

References

Beck E (1983) Frost- und Feuerresistenz tropisch-alpiner Pflanzen. Naturwiss Rundsch 36:105–109

Beck E (1994a) Cold tolerance in tropical alpine plants. In: Rundel PW, Smith AP, Meinzer FC (eds) Tropical alpine environments. Plant form and function. Cambridge University Press, Cambridge, pp 77–110

Beck E (1994b) Turnover and conservation of nutrients in the pachycaul *Senecio keniodendron*. In: Rundel PW, Smith AP, Meinzer FC (eds) Tropical alpine environments. Plant form and function. Cambridge University Press, Cambridge, pp 215–221

Beck E, Scheibe R, Senser M, Müller W (1980) Estimation of leaf and stem growth of unbranched *Senecio keniodendron* trees. Flora 170:68–76

Beck E, Senser M, Scheibe R, Steiger H-M, Pongratz P (1982) Frost avoidance and freezing tolerance in afroalpine "giant-rosette" plants. Plant Cell Environ 5:215–222

Beck E, Scheibe R, Senser M (1983) The vegetation of the Shira Plateau and the western slopes of Kibo (Mt. Kilimanjaro, Tanzania). Phytocoenologia 11:1–30

Beck E, Schulze E-D, Senser M, Scheibe R (1984) Equilibrium freezing of leaf water and extracellular ice formation in afroalpine "giant-rosette" plants. Planta 162:276–282

Carlquist S (1994) Anatomy of tropical alpine plants. In: Rundel PW, Smith AP, Meinzer FC (eds) Tropical alpine environments. Plant form and function. Cambridge University Press, Cambridge, pp 111–128

Goldstein G, Nobel PS (1991) Changes in osmotic pressure and mucilage during low-temperature acclimation of *Opuntia ficus-indica*. Plant Physiol 97:954–961

Goldstein G, Nobel PS (1994) Water relations and low-temperature acclimation for cactus species varying in freezing tolerance. Plant Physiol 104:675–681

Hedberg O (1964a) Features of afroalpine plant ecology. Acta Phytogeogr Suec 49:1–144

Hedberg O (1964b) Etudes écologiques de la flore afroalpine. Bull Soc R Bot Belg 97:5–18

Jones HG (1992) Plants and microclimates, 2nd edn. Cambridge University Press, Cambridge

Keeley JE, Keeley SC (1989) Crassulacean acid metabolism (CAM) in high elevation tropical cactus. Plant Cell Environ 12:331–336

Keeley JE, DeMason DA, Gonzalez R, Markham KR (1994) Sediment-based carbon nutrition in tropical alpine *Isoëtes*. In: Rundel PW, Smith AP, Meinzer FC (eds) Tropical alpine environments. Plant form and function., Cambridge University Press, Cambridge, pp 167–194

Krog JO, Zachariassen KE, Larsen B, Smidsrod O (1979) Thermal buffering in afroalpine plants due to nucleating agent-induced water freezing. Nature 282:300–301

Lang M, Schindler C (1994) The effect of leaf-hairs on blue and red fluorescence emission and on zeaxanthin cycle performance of *Senecio medley* L. J Plant Physiol 144:680–685

Lauer W (1975) Vom Wesen der Tropen. Klimaökologische Studien zum Inhalt und zur Abgrenzung eines irdischen Landschaftsgürtels. Akad Wiss Lit Abh Math Naturwiss Kl (Mainz) 1975, 3:5–52

Lipp CC, Goldstein G, Meinzer FC, Niemczura W (1994) Freezing tolerance and avoidance in high-elevation Hawaiian plants. Plant Cell Environ 17:1035–1044

Medina E, Delgado M (1976) Photosynthesis and night CO_2-fixation in *Echeveria columbiana* Poellnitz. Photosynthetica 10:155–163

Meinzer F, Goldstein G (1985) Some consequences of leaf pubescence in the Andean giant-rosette plant *Espeletia timotensis*. Ecology 66:512–520

Meinzer FC, Goldstein G, Rundel PW (1994) Comparative water relations of tropical alpine plants. In: Rundel PW, Smith AP, Meinzer FC (eds) Tropical alpine environments. Plant form and function. Cambridge University Press, Cambridge, pp 61–76

Melcher PJ, Goldstein G, Meinzer FC, Minyard B, Giambelluca TW, Loope LL (1994) Determinants of thermal balance in the Hawaiian giant rosette plant, *Argyroxiphium sandwicense*. Oecologia 98:412–418

Miller GA (1994) Functional significance of inflorescence pubescence in tropical alpine species of *Puya*. In: Rundel PW, Smith AP, Meinzer FC (eds) Tropical alpine environments. Plant form and function. Cambridge University Press, Cambridge, pp 195–213

Rada F, Goldstein G, Azocar A, Meinzer F (1985) Freezing avoidance in Andean giant rosette plants. Plant Cell Environ 8:501–507

Rehder H (1994) Soil nutrient dynamics in East African alpine ecosystems. In: Rundel PW, Smith AP, Meinzer FC (eds) Tropical alpine environments. Plant form and function. Cambridge University Press, Cambridge, pp 223–228

Reisigl H, Keller R (1987) Alpenpflanzen im Lebensraum. G Fischer, Stuttgart

Rundel PW (1994) Tropical alpine climates. In: Rundel PW, Smith AP, Meinzer FC (eds) Tropical alpine environments. Plant form and function. Cambridge University Press, Cambridge, pp 21–44

Rundel PW, Smith AP, Meinzer FC (eds) (1994a) Tropical alpine environments. Plant form and function. Cambridge University Press, Cambridge

Rundel PW, Meinzer FC, Smith AP (1994b) Tropical alpine ecology:progress and priorities. In: Rundel PW, Smith AP, Meinzer FC (eds) Tropical alpine environments. Plant form and function. Cambridge University Press, Cambridge, pp 355–363

Schuepp PH (1993) Leaf boundary layers. New Phytol 125:477–507

Smith AP (1974) Bud temperature in relation to nyctinastic leaf movement in an Andean giant-rosette plant. Biotropica 6:263–266

Squeo FA, Rada F, Azocar A, Goldstein G (1991) Freezing tolerance and avoidance in high tropical Andean plants:is it equally represented in species with different plant height? Oecologia 86:378–382

Troll C (1943) Die Frostwechselhäufigkeit in den Luft- und Bodenklimaten der Erde. Meteorol Z 60:161–171

Walter H, Breckle S-W (1984) Spezielle Ökologie der tropischen und subtropischen Zonen. G Fischer, Stuttgart

Zhu JJ, Beck E (1991) Water relations of *Pachysandra* leaves during freezing and thawing. Evidence for a negative pressure potential alleviating freeze-dehydration stress. Plant Physiol 97:1146–1153

Subject Index

Printing: Saladruck, Berlin
Binding: Buchbinderei Lüderitz & Bauer, Berlin